The Human Brain during the First Trimester 31- to 33-mm Crown-Rump Lengths

This fifth of 15 short atlases reimagines the classic 5-volume *Atlas of Human Central Nervous System Development*. This volume presents serial sections from specimens between 31 mm and 33 mm with detailed annotations, together with 3D reconstructions. An introduction summarizes human CNS development by using high-resolution photos of methacrylate-embedded rat embryos at a similar stage of development as the human specimens in this volume. The accompanying Glossary gives definitions for all the terms used in this volume and all the others in the *Atlas*.

Key Features

- Classic anatomical atlas
- Detailed labeling of structures in the developing brain offers updated terminology and the identification of unique developmental features, such as, germinal matrices of specific neuronal populations and migratory streams of young neurons
- Appeals to neuroanatomists, developmental biologists, and clinical practitioners
- A valuable reference work on brain development that will be relevant for decades

ATLAS OF
HUMAN CENTRAL NERVOUS SYSTEM DEVELOPMENT
Series

Volume 1: The Human Brain during the First Trimester 3.5- to 4.5-mm Crown-Rump Lengths

Volume 2: The Human Brain during the First Trimester 6.3- to 10.5-mm Crown-Rump Lengths

Volume 3: The Human Brain during the First Trimester 15- to 18-mm Crown-Rump Lengths

Volume 4: The Human Brain during the First Trimester 21- to 23-mm Crown-Rump Lengths

Volume 5: The Human Brain during the First Trimester 31- to 33-mm Crown-Rump Lengths

Volume 6: The Human Brain during the First Trimester 40- to 42-mm Crown-Rump Lengths

Volume 7: The Human Brain during the First Trimester 57- to 60-mm Crown-Rump Lengths

Volume 8: The Human Brain during the Second Trimester 96- to 150-mm Crown-Rump Lengths

Volume 9: The Human Brain during the Second Trimester 160- to 170-mm Crown-Rump Lengths

Volume 10: The Human Brain during the Second Trimester 190- to 210-mm Crown-Rump Lengths

Volume 11: The Human Brain during the Third Trimester 225- to 235-mm Crown-Rump Lengths

Volume 12: The Human Brain during the Third Trimester 260- to 270-mm Crown-Rump Lengths

Volume 13: The Human Brain during the Third Trimester 310- to 350-mm Crown-Rump Lengths

Volume 14: The Spinal Cord during the First Trimester

Volume 15: The Spinal Cord during the Second and Third Trimesters and the Early Postnatal Period

The Human Brain during the First Trimester 31- to 33-mm Crown-Rump Lengths

Atlas of Human Central Nervous System Development, Volume 5

Shirley A. Bayer and Joseph Altman

CRC Press
Taylor & Francis Group
Boca Raton London New York

CRC Press is an imprint of the
Taylor & Francis Group, an **informa** business

First edition published 2023
by CRC Press
6000 Broken Sound Parkway NW, Suite 300, Boca Raton, FL 33487-2742

and by CRC Press
4 Park Square, Milton Park, Abingdon, Oxon, OX14 4RN

CRC Press is an imprint of Taylor & Francis Group, LLC

LCCN no. 2022008216

ISBN: 978-1-032-18333-6 (hbk)
ISBN: 978-1-032-18332-9 (pbk)
ISBN: 978-1-003-27064-5 (ebk)

DOI: 10.1201/9781003270645

Typeset in Times Roman by KnowledgeWorks Global Ltd.

Access the Support Material at: https://www.routledge.com/9781032183336

CONTENTS

ACKNOWLEDGMENTS

We thank the late Dr. William DeMyer, pediatric neurologist at Indiana University Medical Center, for access to his personal library on human CNS development. We also thank the staff of the National Museum of Health and Medicine that were at the Armed Forces Institute of Pathology, Walter Reed Hospital, Washington, D.C. when we collected data in 1995 and 1996: Dr. Adrianne Noe, Director; Archibald J. Fobbs, Curator of the Yakovlev Collection; Elizabeth C. Lockett; and William Discher. We are most grateful to the late Dr. James M. Petras at the Walter Reed Institute of Research who made his darkroom facilities available so that we could develop all the photomicrographs on location rather than in our laboratory in Indiana. Finally, we thank Chuck Crumley, Neha Bhatt, Kara Roberts, Michele Dimont, and Rebecca Condit for expert help during production of the manuscript.

AUTHORS

Shirley A. Bayer received her PhD from Purdue University in 1974 and spent most of her scientific career working with Joseph Altman. She was a professor of biology at Indiana-Purdue University in Indianapolis for several years, where she taught courses in human anatomy and developmental neurobiology while continuing to do research in brain development. Her lengthy publication record of dozens of peer-reviewed scientific journal articles extends back to the mid 1970s. She has co-authored several books and many articles with her late spouse, Joseph Altman. It was her research (published in *Science* in 1982) that proved that new neurons are added to granule cells in the dentate gyrus during adult life, a unique neuronal population that grows. That paper stimulated interest in the dormant field of adult neurogenesis.

Joseph Altman, now deceased, was born in Hungary and migrated with his family via Germany and Australia to the United States. In New York, he became a graduate student in psychology in the laboratory of Hans-Lukas Teuber, earning a PhD in 1959 from New York University. He was a postdoctoral fellow at Columbia University, and later joined the faculty at the Massachusetts Institute of Technology. In 1968, he accepted a position as a professor of biology at Purdue University. During his career, he collaborated closely with Shirley A. Bayer. From the early 1960s to 2016, he published many articles in peer-reviewed journals, books, monographs, and online free books that emphasized developmental processes in brain anatomy and function. His most important discovery was adult neurogenesis, the creation of new neurons in the adult brain. This discovery was made in the early 1960s while he was based at MIT and was largely ignored in favor of the prevailing dogma that neurogenesis is limited to prenatal development. After Dr. Bayer's paper proved that new neurons are adding to granule cells in the hippocampus, his monumental discovery became more accepted. During the 1990s, new researchers "rediscovered" and confirmed his original finding. Adult neurogenesis has recently been proven to occur in the dentate gyrus, olfactory bulb, and striatum through the measurement of Carbon-14—the levels of which changed during nuclear bomb testing throughout the 20th century—in postmortem human brains. Today, many laboratories around the world are continuing to study the importance of adult neurogenesis in brain function. In 2011, Dr. Altman was awarded the Prince of Asturias Award, an annual prize given in Spain by the Prince of Asturias Foundation to individuals, entities, or organizations from around the world who make notable achievements in the sciences, humanities, and public affairs. In 2012, he received the International Prize for Biology, an annual award from the Japan Society for the Promotion of Science (JSPS) for "outstanding contribution to the advancement of research in fundamental biology." This prize is one of the most prestigious honors a scientist can receive. Dr. Altman died in 2016, and Dr. Bayer continues the work they started over 50 years ago. In his honor, she has set up the Altman Prize, awarded each year to an outstanding young researcher in developmental neuroscience by JSPS.

INTRODUCTION

ORGANIZATION OF THE ATLAS

This is the fifth book in the *Atlas of Human Central Nervous System Development* series, 2nd Edition. It deals with human brain development in three normal specimens during the first trimester with crown-rump (CR) lengths from 32- to 33-mm and estimated gestation weeks (GW) from 9.5 to 9.6 (Loughna et al., 2008). These specimens were analyzed in Volume 4 of the 1st Edition (Bayer and Altman, 2006) and all are from the *Carnegie Collection*[1] in the National Museum of Health and Medicine that used to be housed at the Armed Forces Institute of Pathology (AFIP) in Walter Reed Hospital in Washington, D.C. Since the AFIP closed, the National Museum was moved to Silver Springs, MD; this collection is still available for research. C609 is cut in the horizontal plane; C9226 in the frontal plane, and C145 is cut in the sagittal plane. The three section planes in specimens of the same age give a full perspective of the structure of the brain at this time. As in the previous volumes of this Atlas, each specimen is presented in serial grayscale photographs of its Nissl-stained sections showing the brain and surrounding tissues (**Parts II–IV**). The photographs are shown from ventral to dorsal (horizontal specimen), anterior to posterior (frontal specimen), and medial to lateral (sagittal specimen). In the horizontal specimen, the left side of the photo is anterior, right side, posterior and the midline is in the center. The dorsal part of each frontal photo is toward the top of the page, the ventral part at the bottom, and the midline is in the vertical center. In the sagittal specimen, the left side of each photo is anterior, right side posterior, top side dorsal, and bottom side ventral.

---

1. The *Carnegie Collection* (designated by a **C** prefix in the specimen number) started in the Department of Embryology of the Carnegie Institution of Washington. It was led by Franklin P. Mall (1862-1917), George L. Streeter (1873-1948), and George W. Corner (1889-1981). These specimens were collected during a 40–50-year time span and were histologically prepared with a variety of fixatives, embedding media, cutting planes, and histological stains. Early analyses of specimens were published in the early 1900s in *Contributions to Embryology, The Carnegie Institute of Washington* (now archived in the Smithsonian Libraries). O'Rahilly and Müller (1987, 1994) have given overviews of some first trimester specimens in this collection.

PLATE PREPARATION

All sections of a given specimen were photographed at the same magnification. Sections throughout the entire specimen were photographed in serial order with Kodak technical pan black-and-white negative film (#TP442). The film was developed for 6 to 7 minutes in dilution F of Kodak HC-110 developer, stop bath for 30 seconds, Kodak fixer for 5 minutes, Kodak hypo-clearing agent for 1 minute, running water rinse for 10 minutes, and a brief rinse in Kodak photo-flo before drying. The negatives were scanned at 2700 dots per inch (dpi) with a Nikon Coolscan-1000 35-mm negative film scanner attached to a Macintosh PowerMac G3 computer, which had a plug-in driver built into Adobe Photoshop. The negatives were scanned as color positives because that brought out more subtle shades of gray. The original scans were converted to 300 dpi using the non-resampling method for image size. The powerful features of Adobe Photoshop were used to enhance contrast, correct uneven staining, and slightly darken or lighten areas of uneven exposure.

The photos chosen for annotation in **Parts II–IV** are presented as companion plates. The *low-magnification plates* of the horizontal and frontal specimens are designated as **A** and **B** on one set of facing pages. **Part A** on the left shows the full-contrast photo, while **Part B** on the right shows a low-contrast copy with annotations. Plates of the sagittal specimen are designated as **A** through **D** on two sets of facing pages. **Part A** on the left page shows the full-contrast photograph of the brain in the skull without labels. **Part B** on the right page shows low-contrast copies of the same photograph with superimposed outlines of brain parts and labels of major brain ventricles and structures. **Part C** on the second left page shows a full-contrast photo of a slightly larger brain "dissected" from its peripheral structures, except cranial sensory structures have been preserved. **Part D** on the second right page is a low-contrast copy of the photo in C with more detailed labeling in the brain. Several *high-magnification plates* feature enlarged views of the brain to show tissue organization. This type of presentation allows a user to see the entire section and then consult the detailed markup in the low-contrast copy on the facing page, leaving little doubt about what is being identified. The labels themselves are not abbreviated, so the user is not constantly having to consult a list. Different fonts are used to label different classes of structures: the ventricular system is labeled in **CAPITALS**, the neuroepithelium and other germinal zones in **Helvetica bold**, transient structures in ***Times bold italic***, and permanent structures in Times Roman or **Times bold**. Adobe Illustrator was used to superimpose labels and to outline structural details on the low-contrast images. Plates were placed into a book layout using Adobe InDesign. Finally, high-resolution portable document files (pdf) were uploaded to CRC Press/Taylor & Francis websites.

DEVELOPMENT IN SPECIMENS
(CR 31-33-mm)

The specimens in this volume are equivalent to rat embryos on embryonic day (E) 17 based on our morphological matching. E17 rats have a similar appearance to human specimens from 31- to 33-mm crown-rump lengths. Our timetables of neurogenesis used ^{3}H-thymidine dating methods (Bayer and Altman, 1991, 1995, 2012-present; Bayer et al., 1993, 1995) to determine neuronal populations that are being generated in E17 rats; we assume that is comparable to neurogenesis in 31– to 33–mm human specimens (Bayer et al., 1993, 1995; Bayer and Altman, 1995). **Table 1** lists populations being generated throughout the neuraxis: the brainstem, cerebellum, and midbrain tectum (**Table 1A**), the hypothalamus and preoptic area (**Table 1B**), the thalamus and epithalamus (**Table 1C**), the pallidum/striatum, amygdala, and septum (**Table 1D**), and the cerebral cortex, hippocampus, and olfactory structures (**Table 1E**). We use photos of methacrylate-embedded rat embryos on E17 to show the fine details of development in major parts of the brain (**Figs. 1-14**, Bayer, 2013-present) because the preservation of human specimens does not often show great detail.

Figure 1 shows the inferior olive in the medulla and the posterior extramural migratory stream crossing the midline; **Figure 2,** the elaborate precerebellar neuroepithelium in the dorsal medulla; **Figure 3,** the cerebellum with a new feature, the external germinal layer growing beneath the pia. This secondary germinal matrix will produce basket, stellate, and granule cells in the cerebellar cortex. **Figure 4** shows the neuroepithelia in the superior and inferior colliculi as they actively generate neuronal populations there. **Figures 5 to 8** show development in the hypothalamus, thalamus, and epithalamus as many neuronal populations are still being generated there. **Figures 9-12** show highlights of the developing basal ganglia, namely, the striatum, septum, and amygdala. Finally, **Figure 13** shows highlights of the developing cerebral cortex. Readers are encouraged to consult the definitions of annotations in the glossary (Bayer, 2022) that accompanies the *Atlas*.

Table 1A: Neurogenesis by Region	
REGION and NEURAL POPULATION	CROWN-RUMP LENGTH 31-33 mm
PRECEREBELLAR NUCLEI	
Pontine reticular nucleus	●
Pontine nuclei	● ●
MESENCEPHALIC TEGMENTUM/ISTHMUS	
Dorsal central gray	● ●
SUPERIOR COLLICULUS	
stratum album	●
stratum griseum profundum	●
stratum lemnisci	●
stratum griseum intermediate	●
stratum opticum	●
stratum griseum superficial	● ●
stratum zonale	●
INFERIOR COLLICULUS	
Anterolateral	● ●
Posterolateral	● ●
Anterior intermediate	● ●
Posterior intermediate	● ●
Anteromedial	● ●
Posteromedial	●

Table 1A. Neural populations in the precerebellar nuclei, mesencephalic tegmentum, superior colliculus, and inferior colliculus that are being generated in rats on Embryonic day (E) 17 (comparable to humans at CR 31– to 33–mm). *Green dots* indicate the amount of neurogenesis occurring: one dot=<15%; two dots=15-90%. This same dot notation is used for all of the remaining parts (**B-E**) of **Table 1**.

Table 1B: Neurogenesis by Region

REGION and NEURAL POPULATION	CROWN-RUMP LENGTH 31-33 mm
PREOPTIC AREA/ HYPOTHALAMUS	
Medial preoptic area	•
Medial preoptic nucleus	••
Sexually dimorphic nucleus	••
Periventricular preoptic nucleus	••
Median preoptic nucleus	•
Ventromedial nucleus	•
Dorsomedial nucleus	••
Arcuate nucleus	••
Suprachiasmatic nucleus	••
Supramammillary nucleus	••
Tuberomammillary nucleus	••
Medial mammillary n. (ventral)	•

Table 1C: Neurogenesis by Region

REGION and NEURAL POPULATION	CROWN-RUMP LENGTH 31-33 mm
THALAMUS/EPITHALAMUS	
Anterodorsal	•
Anteroventral	•
Anteromedial	••
Medial Dorsal	••
Paraventricular	••
Paratenial	••
Reuniens	••
Rhomboid	••
Medial habenula	••

Table 1D: Neurogenesis by Region

REGION and NEURAL POPULATION	CROWN-RUMP LENGTH 31-33 mm
PALLIDUM AND STRIATUM	
Olfactory tubercle (small neurons)	••
Caudate and putamen	•
Nucleus accumbens	•
Islands of Calleja	•
AMYGDALA	
Central nucleus	•
Intercalated masses	••
Amygdalo-hippocampal area	••
Anterior cortical nucleus	•
Posterior cortical nucleus	•
Basomedial nucleus	•
Basolateral nucleus	•
Lateral nucleus	•
Bed n. stria terminalis (anterior)	••
Bed n. stria terminalis (preoptic continuation)	••
SEPTUM	
Medial nucleus	•
Diagonal band (vertical limb)	••
Lateral nucleus	••
Bed nucleus of the anterior commissure	•

Table 1E: Neurogenesis by Region

REGION and NEURAL POPULATION	CROWN-RUMP LENGTH 31-33 mm
NEOCORTEX and LIMBIC CORTEX	
Cajal-Retzius neurons	•
Layer V	••
Layer VI	••
Layera IV-II	••
Subplate VII	•
OLFACTORY CORTEX	
Layer II (anterior)	••
Layer II (posterior)	••
Layers III-IV (anterior)	•
HIPPOCAMPAL REGION	
Entorhinal cortex Layer II	••
Entorhinal cortex Layer III	••
Entorhinal cortex Layer IV	••
Entorhinal cortex Layers V-VI	•
Subiculum (deep)	••
Subiculum (superficial)	••
Ammon's Horn CA1	••
Ammon's Horn CA3	••
Ammon's Horn CA4	••
OLFACTORY BULB	
Mitral cells (main bulb)	•
Internal tufted cells (main bulb)	••
External tufted cells (main bulb)	•
ANTERIOR OLFACTORY NUCLEUS	
Pars externa	••
AON proper	•

THE LOWER MEDULLA IN AN E17 RAT EMBRYO

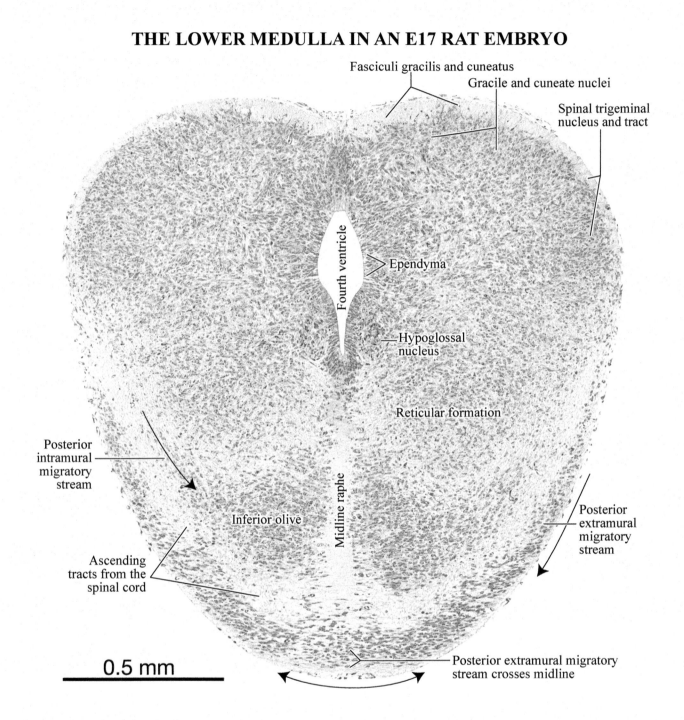

Figure 1. Transverse section of the lower medulla in an E17 rat embryo at a similar stage as human specimens in this volume. The two posterior migratory streams (*arrows*) of precerebellar nuclear neurons are clearly visible. The intramural stream takes a trajectory through the medullary parenchyma and heads for the inferior olive. The extramural stream runs beneath the pia meninx on the outer edge of the medulla. Neurons cross the midline and will settle on the opposite side in the lateral reticular and external cuneate precerebellar nuclei. (3μ methacrylate section, toluidine blue stain). Source: braindevelopmentmaps.org (E17 horizontal archive)

THE PRECEREBELLAR NEUROEPITHELIUM
IN THE UPPER MEDULLA IN AN E17 RAT EMBRYO

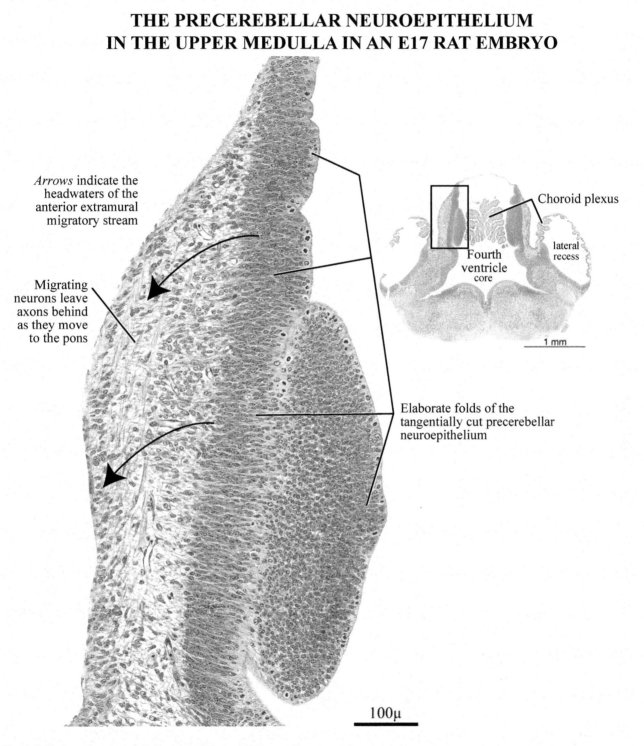

Arrows indicate the headwaters of the anterior extramural migratory stream

Migrating neurons leave axons behind as they move to the pons

Choroid plexus

lateral recess

Fourth ventricle
core

1 mm

Elaborate folds of the tangentially cut precerebellar neuroepithelium

100μ

Figure 2. Horizontal section of the dorsal medulla in an E17 rat embryo at a similar stage as human specimens in this volume. The elaborate precerebellar neuroepithelium is shown in detail. This is the germinal source of the pontine reticular and pontine precerebellar nuclei that are currently being generated. Young neurons migrate out of the neuroepithelium to form the anterior extramural migratory stream that travels beneath the pia to the pons. (3μ methacrylate section, toluidine blue stain). Source: braindevelopmentmaps.org (E17 horizontal archive)

6

THE CEREBELLUM IN AN E17 RAT EMBRYO

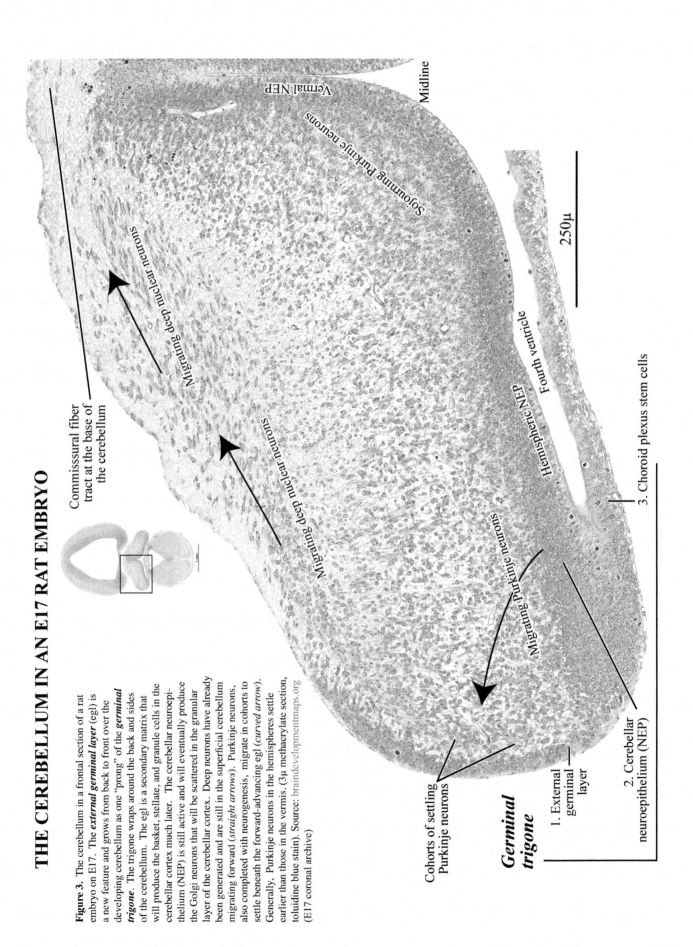

Figure 3. The cerebellum in a frontal section of a rat embryo on E17. The *external germinal layer* (egl) is a new feature and grows from back to front over the developing cerebellum as one "prong" of the *germinal trigone*. The trigone wraps around the back and sides of the cerebellum. The egl is a secondary matrix that will produce the basket, stellate, and granule cells in the cerebellar cortex much later. The cerebellar neuroepithelium (NEP) is still active and will eventually produce the Golgi neurons that will be scattered in the granular layer of the cerebellar cortex. Deep neurons have already been generated and are still in the superficial cerebellum migrating forward (*straight arrows*). Purkinje neurons, also completed with neurogenesis, migrate in cohorts to settle beneath the forward-advancing egl (*curved arrow*). Generally, Purkinje neurons in the hemispheres settle earlier than those in the vermis. (3μ methacrylate section, toluidine blue stain). Source: braindevelopmentmaps.org (E17 coronal archive)

Commisssural fiber tract at the base of the cerebellum

Vermal NEP

Sojourning Purkinje neurons

Midline

Migrating deep nuclear neurons

Migrating deep nuclear neurons

Hemispheric NEP

Fourth ventricle

250μ

Migrating Purkinje neurons

3. Choroid plexus stem cells

Cohorts of settling Purkinje neurons

Germinal trigone

1. External germinal layer

2. Cerebellar neuroepithelium (NEP)

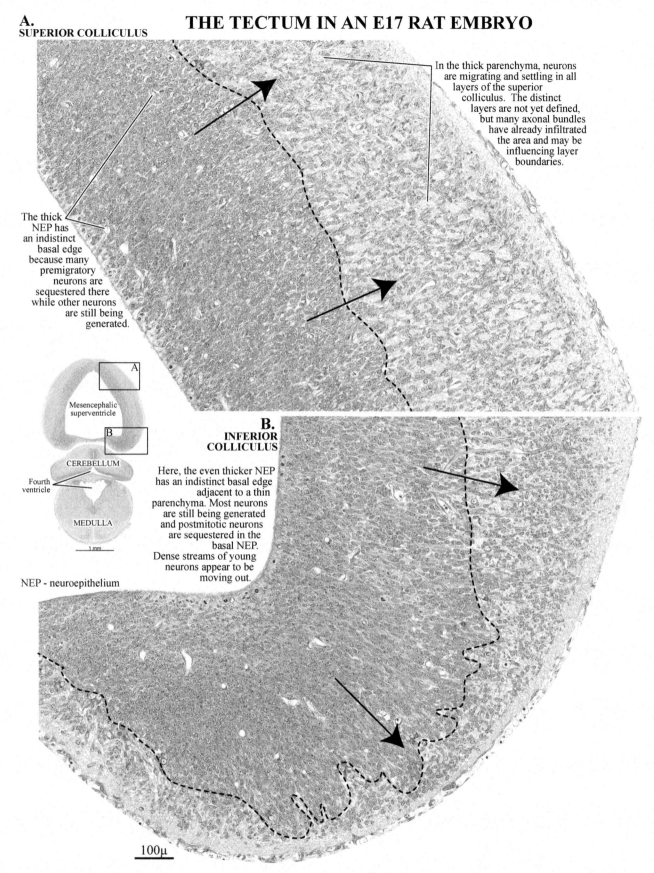

A.
SUPERIOR COLLICULUS

THE TECTUM IN AN E17 RAT EMBRYO

In the thick parenchyma, neurons are migrating and settling in all layers of the superior colliculus. The distinct layers are not yet defined, but many axonal bundles have already infiltrated the area and may be influencing layer boundaries.

The thick NEP has an indistinct basal edge because many premigratory neurons are sequestered there while other neurons are still being generated.

Mesencephalic superventricle

CEREBELLUM

Fourth ventricle

MEDULLA

1 mm

NEP - neuroepithelium

B.
INFERIOR COLLICULUS

Here, the even thicker NEP has an indistinct basal edge adjacent to a thin parenchyma. Most neurons are still being generated and postmitotic neurons are sequestered in the basal NEP. Dense streams of young neurons appear to be moving out.

100µ

Figure 4. The same frontal section as in **Fig. 3** showing detail of the tectal neuroepithelia (NEP) for the superior colliculus (**A**) and inferior colliculus (**B**) in an E17 rat embryo. *Arrows* indicate neurons migrating out of the NEPs. (3µ methacrylate section, toluidine blue stain). Source: braindevelopmentmaps.org (E17 coronal archive)

THE MIDLEVEL FOREBRAIN IN AN E17 RAT EMBRYO

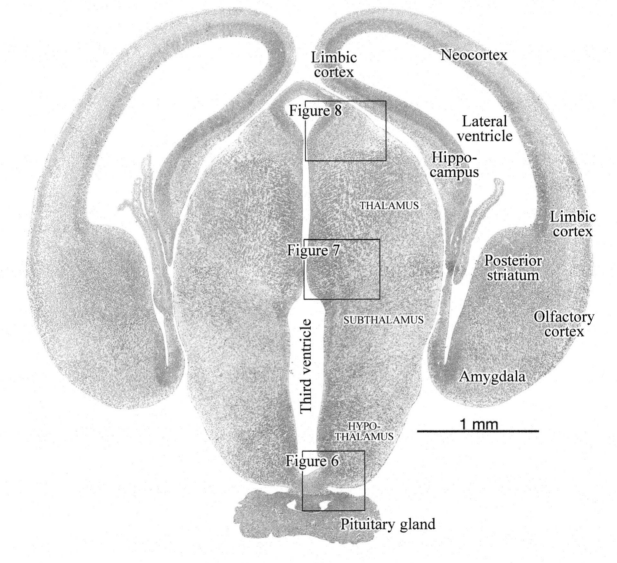

Figure 5. Frontal section of the diencephalon and cerebral cortex in an E17 rat embryo at the same stage of development as the human specimens in this volume. The boxes indicate the locations of the detailed photos of the hypothalamus (**Fig. 6**), intermediate lobule of the thalamus (**Fig. 7**), and the epithalamus (**Fig. 8**). (3µ methacrylate section, toluidine blue stain) Source: braindevelopmentmaps.org (E17 coronal archive)

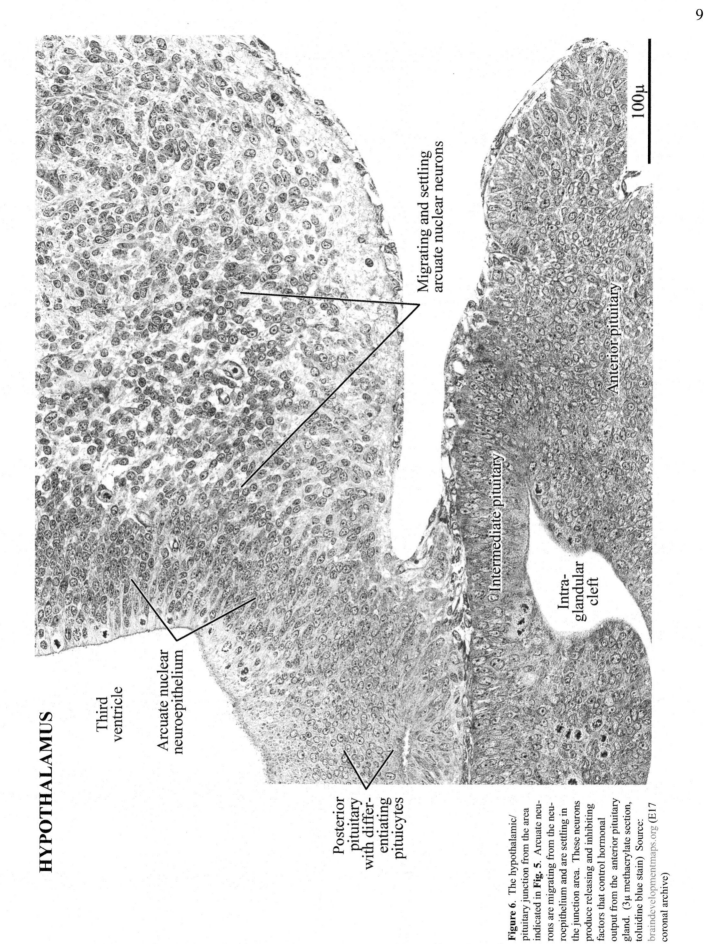

HYPOTHALAMUS

Third ventricle

Arcuate nuclear neuroepithelium

Migrating and settling arcuate nuclear neurons

Posterior pituitary with differentiating pituicytes

Intermediate pituitary

Anterior pituitary

Intra-glandular cleft

100μ

Figure 6. The hypothalamic/ pituitary junction from the area indicated in **Fig. 5**. Arcuate neurons are migrating from the neuroepithelium and are settling in the junction area. These neurons produce releasing and inhibiting factors that control hormonal output from the anterior pituitary gland. (3μ methacrylate section, toluidine blue stain) Source: braindevelopmentmaps.org (E17 coronal archive)

THALAMUS

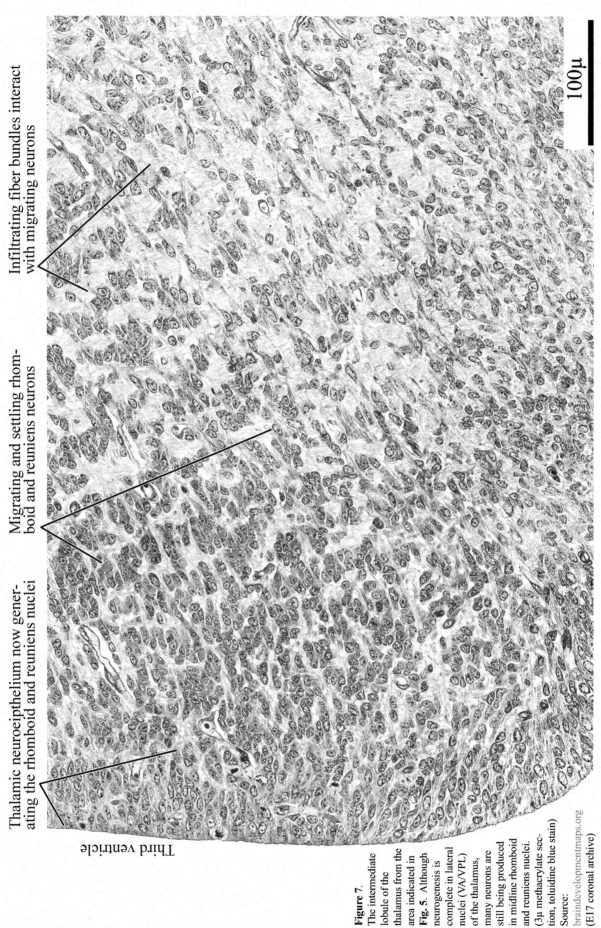

Infiltrating fiber bundles interact with migrating neurons

Migrating and settling rhomboid and reuniens neurons

Thalamic neuroeipthelium now generating the rhomboid and reuniens nuclei

Third ventricle

Figure 7.
The intermediate lobule of the thalamus from the area indicated in **Fig. 5.** Although neurogenesis is complete in lateral nuclei (VA/VPL) of the thalamus, many neurons are still being produced in midline rhomboid and reuniens nuclei. (3μ methacrylate section, toluidine blue stain)
Source:
braindevelopmentmaps.org
(E17 coronal archive)

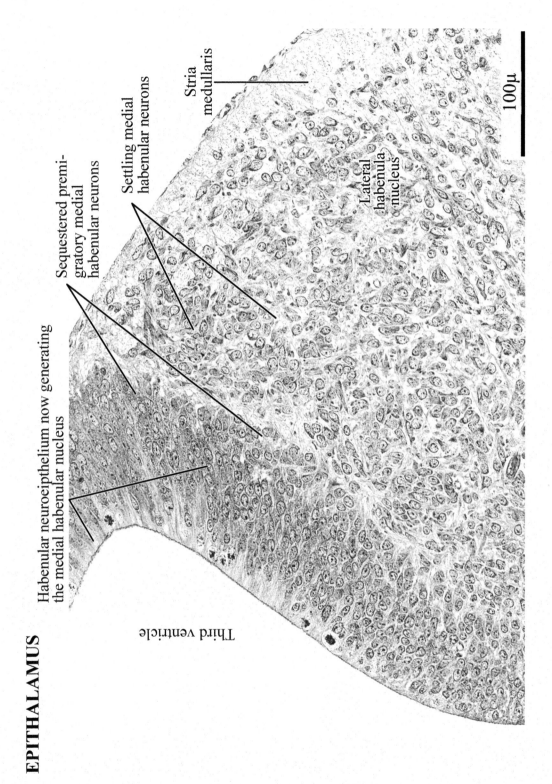

EPITHALAMUS

Stria medullaris

Settling medial habenular neurons

Sequestered premigratory medial habenular neurons

Habenular neuroeipthelium now generating the medial habenular nucleus

Lateral habenula nucleus

Third ventricle

100μ

Figure 8. The epithalamus from the area indicated in **Fig. 5.** The lateral habenula nucleus has been generated and is settling, but the medial habenular nucleus is still being generated. (3μ methacrylate section, toluidine blue stain) Source: braindevelopmentmaps.org (E17 coronal archive)

12

THE TELENCEPHALON IN AN E17 RAT EMBRYO

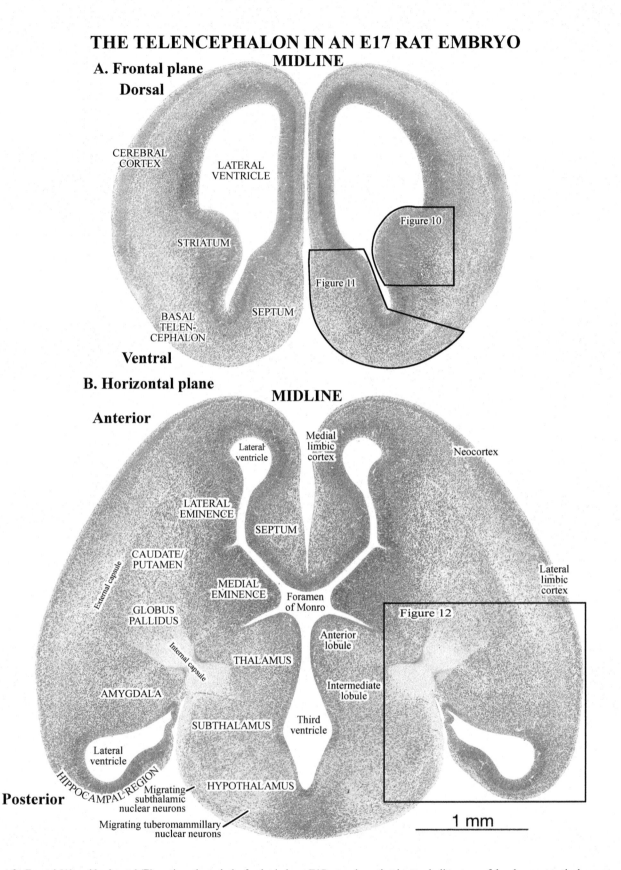

A. Frontal plane

MIDLINE

Dorsal

CEREBRAL CORTEX

LATERAL VENTRICLE

Figure 10

Figure 11

STRIATUM

SEPTUM

BASAL TELEN-CEPHALON

Ventral

B. Horizontal plane

MIDLINE

Anterior

Lateral ventricle

Medial limbic cortex

Neocortex

LATERAL EMINENCE

SEPTUM

CAUDATE/ PUTAMEN

External capsule

MEDIAL EMINENCE

Foramen of Monro

Lateral limbic cortex

GLOBUS PALLIDUS

Figure 12

Internal capsule

Anterior lobule

THALAMUS

AMYGDALA

Intermediate lobule

SUBTHALAMUS

Third ventricle

Lateral ventricle

HIPPOCAMPAL REGION

Migrating subthalamic nuclear neurons

HYPOTHALAMUS

Posterior

Migrating tuberomammillary nuclear neurons

1 mm

Figure 9. Frontal (**A**) and horizontal (**B**) sections through the forebrain in an E17 rat embryo that is at a similar stage of development to the human specimens in this volume. The boxes indicate the locations of the high-magnification photos in **Figs. 10-12**. (3µ methacrylate sections, toluidine blue stain) Sources: braindevelopmentmaps.org (E17 coronal and horizontal archives)

THE ANTEROLATERAL GANGLIONIC EMINENCE

The indistinct basal edge of the
neuroepithelium (NEP) blends with
the thick subventricular zone (SVZ).

Cortical NEP   Cortical SVZ

Lateral
migratory
stream
from the
cerebral
cortex

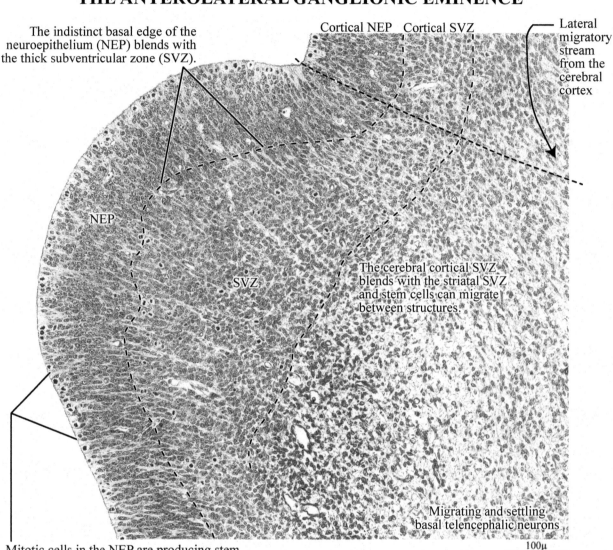

NEP

SVZ

The cerebral cortical SVZ
blends with the striatal SVZ
and stem cells can migrate
between structures.

Migrating and settling
basal telencephalic neurons

100μ

Mitotic cells in the NEP are producing stem
cells that will migrate to the SVZ and generate
medium spiny neurons in the striatum.

Figure 10. Frontal section of the anterolateral ganglionic eminence from the area indicated in (**Fig. 9A**) showing the neuroepithelium flanked by a thick subventricular zone. Note that the anterolateral eminence is continuous with the germinal zones and parenchymal layers of the cerebral cortex. (3μ methacrylate section, toluidine blue stain) Source: braindevelopmentmaps.org (E17 coronal and horizontal archives)

THE SEPTUM, NUCLEUS ACCUMBENS, AND MEDIAL BASAL TELENCEPHALON

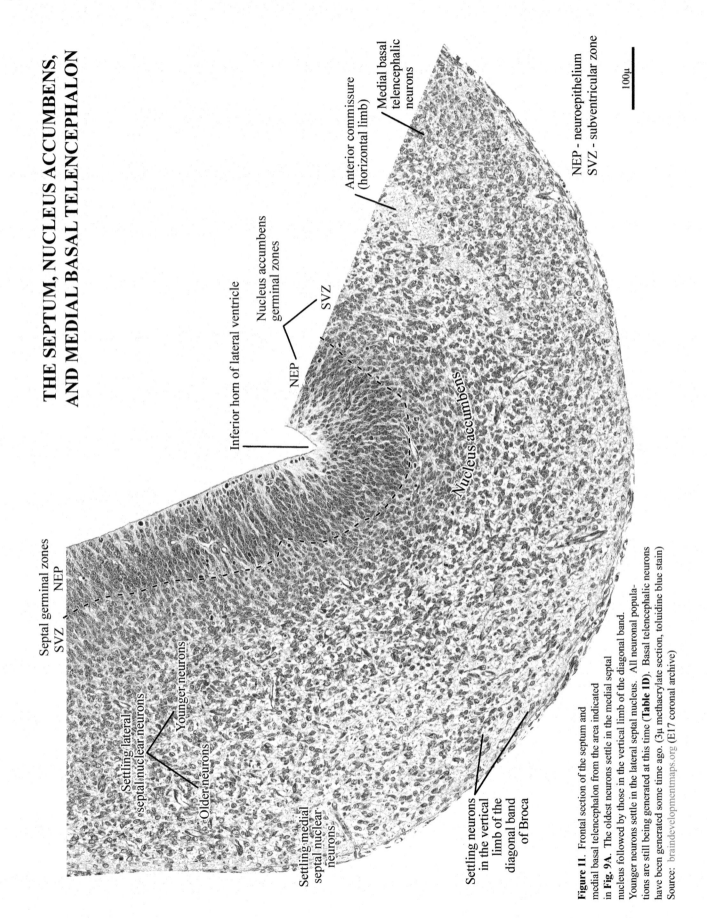

100μ

Septal germinal zones
SVZ   NEP

Settling lateral
septal nuclear neurons

Younger neurons

Older neurons

Settling medial
septal nuclear
neurons

Settling neurons
in the vertical
limb of the
diagonal band
of Broca

Inferior horn of lateral ventricle

Nucleus accumbens
germinal zones

NEP

SVZ

Nucleus accumbens

Anterior commissure
(horizontal limb)

Medial basal
telencephalic
neurons

Medial basal telencephalic neurons

NEP - neuroepithelium
SVZ - subventricular zone

Figure 11. Frontal section of the septum and medial basal telencephalon from the area indicated in **Fig. 9A.** The oldest neurons settle in the medial septal nucleus followed by those in the vertical limb of the diagonal band. Younger neurons settle in the lateral septal nucleus. All neuronal populations are still being generated at this time (**Table 1D**). Basal telencephalic neurons have been generated some time ago. (3μ methacrylate section, toluidine blue stain) Source: braindevelopmentmaps.org (E17 coronal archive)

THE POSTEROLATERAL PART OF THE FOREBRAIN

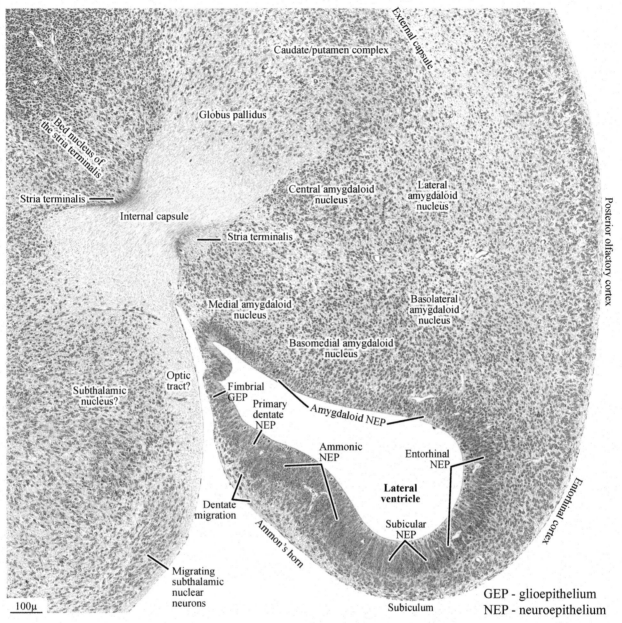

Figure 12. Horizontal section through the amygdala, hippocampal region, and lateral cerebral cortex from the area indicated in **Fig. 9B**.

This section was chosen because it clearly shows the internal capsule bridging the thalamus/subthalamus to the telencephalon. Note that the stria terminalis wraps around the internal capsule anteriorly and posteriorly; anteriorly, it is connected to the bed nucleus of the stria terminalis, posteriorly it enters the amygdala. The amygdaloid NEP is still generating neurons throughout most of its nuclei (**Table 1D**), although some neurons probably migrate into the lateral and basolateral nuclei from the lateral migratory stream of the cerebral cortex. The same is true of the posterior olfactory cortex. Migrating cells from the cortex are perpendicular to the section plane, so they are not clear and are not indicated (*see* **Fig. 10** instead).

The hippocampal region wraps around the medial, posterior and lateral parts of the lateral ventricle. The various NEPs are in neurogenetic stages and are generating neurons for nearly all populations (**Table 1E**). More cells are outside the entorhinal NEP because this part of the region contains the oldest neurons. The subiculum has many neurons outside its NEP, generally destined for its deeper layers. Only a few neurons are outside the Ammonic NEP, probably the oldest in field CA3. A new feature in the region is the primary dentate NEP that shows some cells migrating out to accumulate in the extra space beneath the pia where the dentate gyrus will form. The dentate migration contains some of the oldest dentate granule cells and also progenitors that will establish a dispersed germinal matrix in the dentate hilus and generate granule cells now and into adulthood.

(3μ methacrylate section, toluidine blue stain) Source: braindevelopmentmaps.org (E17 horizontal archive)

16

THE CEREBRAL CORTEX

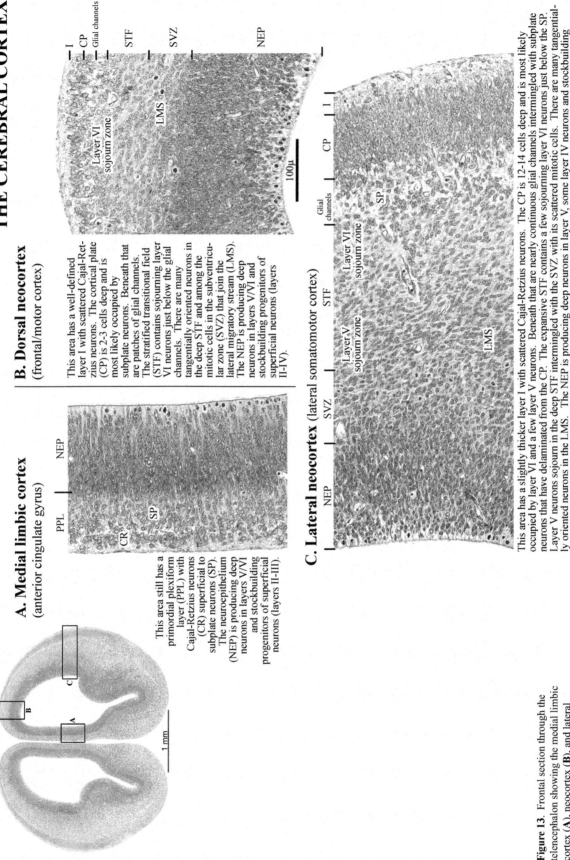

A. Medial limbic cortex (anterior cingulate gyrus)

This area still has a primordial plexiform layer (PPL) with Cajal-Retzius neurons (CR) superficial to subplate neurons (SP). The neuroepithelium (NEP) is producing deep neurons in layers V/VI and stockbuilding progenitors of superficial neurons (layers II-III).

B. Dorsal neocortex (frontal/motor cortex)

This area has a well-defined layer I with scattered Cajal-Retzius neurons. The cortical plate (CP) is 2-3 cells deep and is most likely occupied by subplate neurons. Beneath that are patches of glial channels. The stratified transitional field (STF) contains sojourning layer VI neurons just below the glial channels. There are many tangentially oriented neurons in the deep STF and among the mitotic cells in the subventricular zone (SVZ) that join the lateral migratory stream (LMS). The NEP is producing deep neurons in layers V/VI and stockbuilding progenitors of superficial neurons (layers II-IV).

C. Lateral neocortex (lateral somatomotor cortex)

This area has a slightly thicker layer I with scattered Cajal-Retzius neurons. The CP is 12-14 cells deep and is most likely occupied by layer VI and a few layer V neurons. Beneath that are nearly continuous glial channels intermingled with subplate neurons that have delaminated from the CP. The expansive STF contains a few sojourning layer VI neurons just below the SP. Layer V neurons sojourn in the deep STF intermingled with the SVZ with its scattered mitotic cells. There are many tangentially oriented neurons in the LMS. The NEP is producing deep neurons in layer V; some layer IV neurons and stockbuilding progenitors of superficial neurons (layers III-II). Some of these progenitors migrate to the SVZ to produce neurons there. In rats the various subdivisions of the STF are only seen with ³H-thymidine autoradiography, but the layers are visible in light microscopy in human fetuses during the late first and second trimesters.

Figure 13. Frontal section through the telencephalon showing the medial limbic cortex (**A**), neocortex (**B**), and lateral neocortex (**C**). (3µ methacrylate section, toluidine blue stain) Source: braindevelopmentmaps.org (E17 coronal archive)

REFERENCES

Bayer, SA (2013-present) www.braindevelopmentmaps.org (This website is an image database of methacrylate-embedded normal rat embryos and paraffin-embedded rat embryos exposed to ^{3}H-Thymidine.)

Bayer, SA (2022) *Glossary to Accompany Atlas of Human Central Nervous System Development (Second Edition)* Laboratory of Developmental Neurobiology, Ocala, FL.

Bayer SA, Altman J (1991) *Neocortical Development*, Raven Press, New York.

Bayer SA, Altman J, Russo RJ, Zhang X (1993) Timetables of neurogenesis in the human brain based on experimentally determined patterns in the rat. *Neurotoxicology* **14**: 83-144.

Bayer SA, Altman J, Russo RJ, Zhang X (1995) Embryology. In: *Pediatric Neuropathology*, Serge Duckett, Ed. Williams and Wilkins, pp. 54-107.

Bayer SA, Altman J (1995) Development: Some principles of neurogenesis, neuronal migration and neural circuit formation. In: *The Rat Nervous System*, 2nd Edition, George Paxinos, Ed. Academic Press, Orlando, Florida., pp. 1079-1098.

Bayer SA, Altman J (2006) *Atlas of Human Central Nervous System Development* (First Edition), Volume 4, CRC Press.

Bayer SA, Altman J (2012-present) www.neurondevelopment.org (This website has downloadable pdf files of our scientific papers on rat brain development grouped by subject.)

Hochstetter F (1919) *Beiträge zur Entwicklungsgeschichte des menschlichen Gehirns*. Vol. 1. Leipzig und Wien: Deuticke.

Loughna P, Citty L, Evans T, Chudleigh T (2009) Fetal size and dating: Charts recommended for clinical obstetric practice, *Ultrasound*, 17:161-167.

O'Rahilly R, Müller F. (1987) *Developmental Stages in Human Embryos, Carnegie Institution of Washington*, Publication 637.

O'Rahilly R; Müller F. (1994) *The Embryonic Human Brain*, Wiley-Liss, New York.

PART II: C609
CR 32 mm (GW 9.6)
Horizontal

Specimen C609 from the Carnegie collection is a normal female fetus with a crown-rump length (CR) of 32 mm that was collected in 1916, and is estimated to be in gestational week (GW) 9.6. The entire fetus was embedded in paraffin, cut transversely in 50-μm-thick sections, and stained with aluminum cochineal. Since there is no photograph of this brain before it was embedded and cut, a specimen from Hochstetter (1919) that is only partially comparable to C609 has been modified to show the approximate section plane and external features of the brain at GW9.6 (**Figure 14**). Like most of the specimens in this volume, the sections are not cut exactly in one plane; C609's cortex is cut midway between coronal and horizontal planes, and is presented as a "horizontal" brain. The C609 section planes through the cortex and brainstem are not at the same angle when transferred to Hochstetter's CR27 mm specimen. Instead, brainstem planes of section appear to fan upward and downward from sections in the cortex apparently around a fulcrum centering in the invagination of the medullary velum overlying the rhombencephalic superventricle. We interpret this to indicate that the brain flexures are more loosely folded in the Hochstetter specimen than in C609. But it is difficult to determine how the brainstem is folded in C609 to make the section planes line up with those in the cortex. Photographs of 23 sections are illustrated at low magnification in **Plates 1-10**. High-magnification views of different areas of the brain are shown in **Plates 11-16**. To maximize the image size within the page space, all of C609's sections are rotated 90 degrees (landscape orientation). The anterior part of each section is on the left (page bottom), and the posterior part of each section is on the right (page top).

C609 is similar to the other GW9.6/9.5 specimens in this volume and shows brain structures in a horizontal perspective. The telencephalic and rhombencephalic *superventricles* are obvious, along with the slit-like diencephalic and mesencephalic superventricles. The parenchyma, the area between the superficial border of the *neuroepithelium (NEP) / subventricular zone (SVZ)* and the pial membrane, is the region where neurons migrate, settle, and differentiate. The parenchyma is thick and bordered by a thin NEP in the medulla, pons, and midbrain tegmentum without surrounding dense sojourn zones. Most neurons have been generated here, few are migrating, and most are settled and differentiating. The two exceptions seen in C9226 are also seen in C145. First, presumptive facial motor neurons are clumped near the pontomedullary trench and some are migrating toward their ventral pontine/medullary settling sites. Second, the thicker *precerebellar neuroepithelium* in the medulla is generating predominantly pontine gray neurons; many precerebellar neurons are migrating in the *anterior and posterior extramural migratory streams*. The cerebellar NEP is thicker than that in the pons and medulla, and the cerebellar parenchyma has a dense Purkinje cell sojourn zone below the mass of earlier-generated deep neurons; the *external germinal layer (egl)* is rudimentary but definitely there. The mesencephalic tectal NEP is thick adjacent to a thin parenchyma that contains dense sojourning and migrating neurons; substantial neurogenesis is ongoing in both the superior and inferior colliculi in the midbrain tectum.

Overall, the thalamus is relatively much larger in human embryos than rat embryos at a similar stage (*see* **Part 1**), correlating with the enormous growth of the cerebral cortex at later stages. The diencephalic NEP and thick parenchyma are filled with dense zones of sojourning and migrating neurons. Although many diencephalic neurons have been generated by GW9.6, most of them are still migrating and few have settled. In spite of that, there are large accumulations of fibers (possibly input from the medial lemniscus) in the thalamus and many fibers appear to be exiting from the lateral thalamus to enter the internal capsule, indicating that young thalamic neurons grow axons toward the cerebral cortex as they are migrating and settling. Another prominent feature is the migrating subthalamic nucleus neurons from the mammillary hypothalamic NEP to the subthalamic nucleus.

Within the telencephalon, the cerebral cortex has a thick NEP and a parenchyma that is thicker ventrolaterally than dorsomedially—a prominent neurogenetic and morphogenetic gradient. The *stratified transitional field (STF)* contains *STF1* and *STF5* only in ventrolateral areas. The basal telencephalic NEP/SVZ and parenchyma are both thick because there are large early-generated neuronal populations (for example, globus pallidus and substantia innominata) and massive late-generated populations (striatal neurons in the caudate and putamen). The neurons settling in the basal telencephalon at GW9.6 are the early-generated populations. Nearly all striatal neurons have yet to be generated, but their progenitors are stockbuilding in the massive subventricular zone.

GW9.6 "HORIZONTAL" SECTION PLANES

C609's cutting angle in the cerebral cortex rotates 45° counterclockwise from the true coronal plane (90°), exactly between true horizontal and true coronal.

This brain is less mature (CR 27 mm) with more loosely folded flexures than C609's brain (CR 32 mm). That is why cutting planes in C609's brainstem differ considerably from those in the cortex, many are horizontal. In the illustrated sections on the following pages, the anterior part of each section (left side) is dorsal to the posterior part (right side).

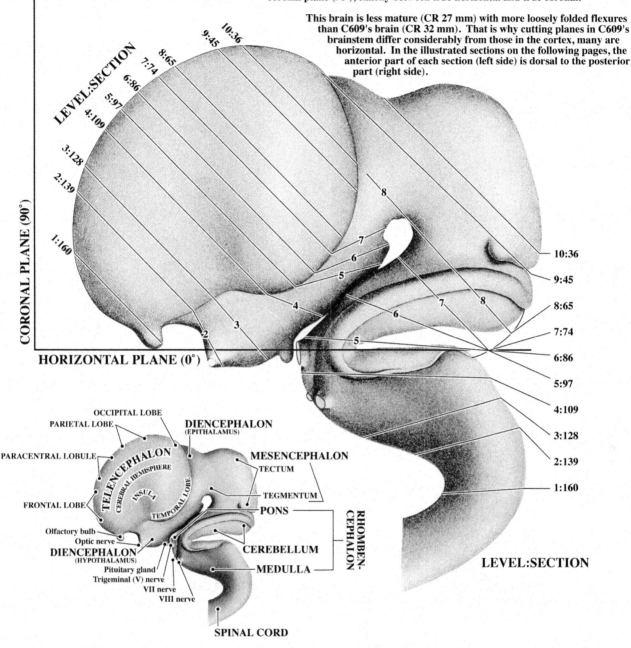

LEVEL:SECTION

LEVEL:SECTION
10:36
9:45
8:65
7:74
6:86
5:97
4:109
3:128
2:139
1:160

CORONAL PLANE (90°)

HORIZONTAL PLANE (0°)

OCCIPITAL LOBE
PARIETAL LOBE
PARACENTRAL LOBULE
TELENCEPHALON
CEREBRAL HEMISPHERE
INSULA
TEMPORAL LOBE
FRONTAL LOBE
Olfactory bulb
Optic nerve
DIENCEPHALON (HYPOTHALAMUS)
Pituitary gland
Trigeminal (V) nerve
VII nerve
VIII nerve
SPINAL CORD

DIENCEPHALON (EPITHALAMUS)
MESENCEPHALON
TECTUM
TEGMENTUM
PONS
CEREBELLUM
MEDULLA
RHOMBEN-CEPHALON

Figure 14. The lateral view of the brain and upper cervical spinal cord from a specimen with a crown-rump length of 27 mm (modified from Figure 37, Table VII, Hochstetter, 1919) serves to show the approximate locations and cutting angles of the illustrated sections of C609 in the following pages. The small inset identifies the major structural features. The lines in the cerebellum and lateral edges of the pons and medulla are the cut edges of the medullary velum.

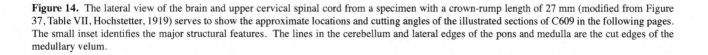

PLATE 1A
CR 32 mm, GW 9.6, C609
Horizontal Section 160

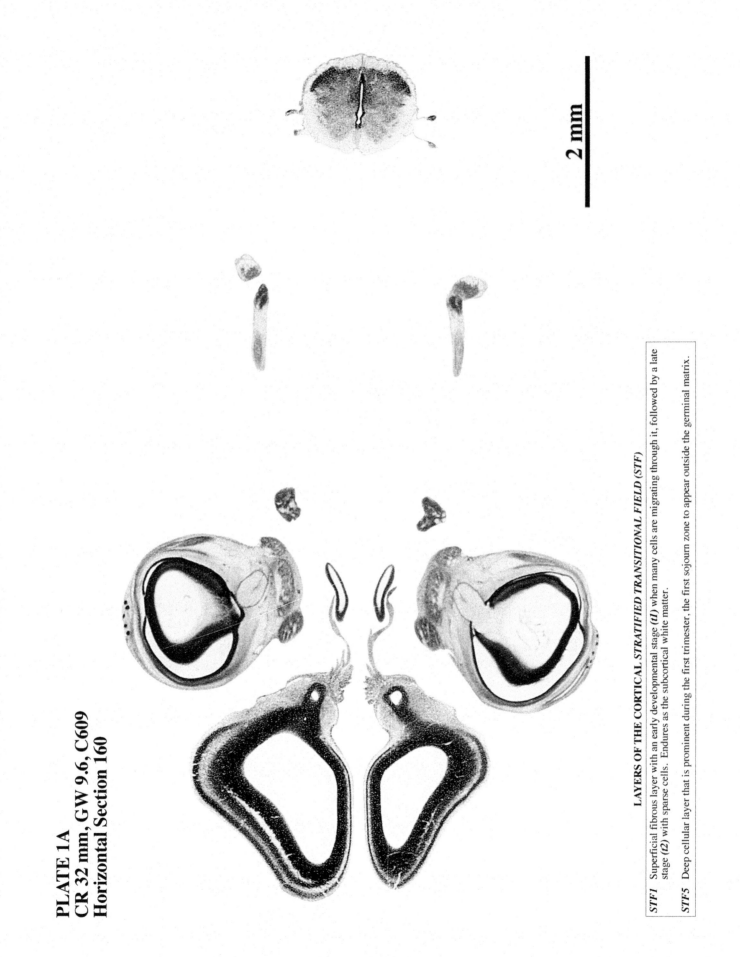

2 mm

LAYERS OF THE CORTICAL *STRATIFIED TRANSITIONAL FIELD (STF)*

STF1 Superficial fibrous layer with an early developmental stage (*t1*) when many cells are migrating through it, followed by a late stage (*t2*) with sparse cells. Endures as the subcortical white matter.

STF5 Deep cellular layer that is prominent during the first trimester, the first sojourn zone to appear outside the germinal matrix.

PLATE 1B

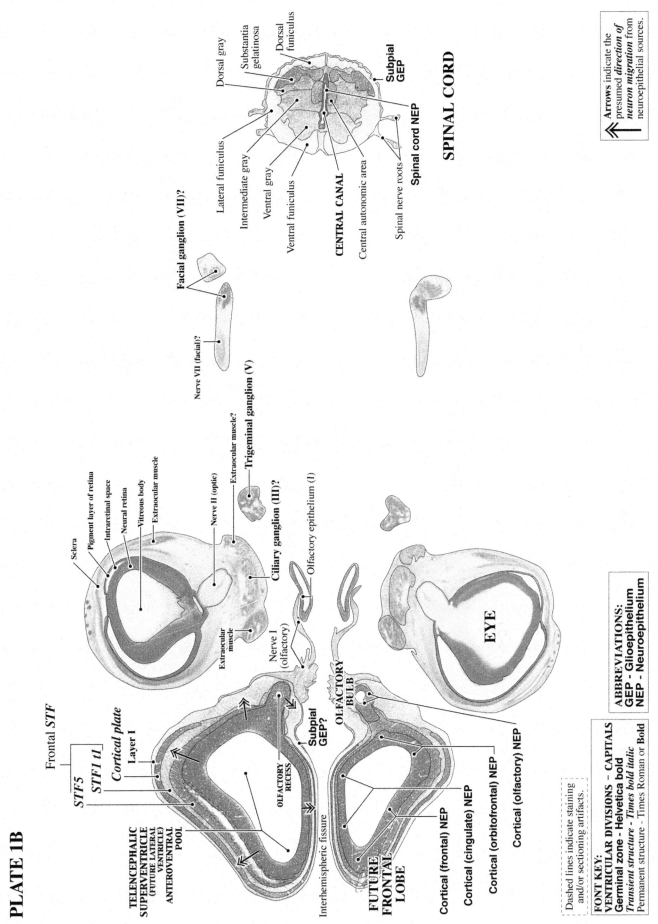

21

Frontal *STF*

STF5
STF1 t1
Cortical plate
Layer I

**TELENCEPHALIC
SUPERVENTRICLE**
(FUTURE LATERAL
VENTRICLE)
ANTEROVENTRAL
POOL

Interhemispheric fissure

OLFACTORY
RECESS

**Subpial
GEP?**

**OLFACTORY
BULB**

**FUTURE
FRONTAL
LOBE**

Cortical (frontal) NEP
Cortical (cingulate) NEP
Cortical (orbitofrontal) NEP
Cortical (olfactory) NEP

Sclera
Pigment layer of retina
Intraretinal space
Neural retina
Vitreous body
Extraocular muscle
Extraocular
muscle
Nerve I
(olfactory)
Nerve II (optic)
Extraocular muscle?
Trigeminal ganglion (V)
Ciliary ganglion (III)?
Olfactory epithelium (I)

Facial ganglion (VII)?
Nerve VII (facial)?

EYE

Dorsal gray
Substantia
gelatinosa
Dorsal
funiculus
**Subpial
GEP**
Lateral funiculus
Intermediate gray
Ventral gray
Ventral funiculus
CENTRAL CANAL
Central autonomic area
Spinal nerve roots
Spinal cord NEP

SPINAL CORD

Arrows indicate the
presumed *direction of
neuron migration* from
neuroepithelial sources.

ABBREVIATIONS:
GEP - Glioepithelium
NEP - Neuroepithelium

Dashed lines indicate staining
and/or sectioning artifacts.

FONT KEY:
VENTRICULAR DIVISIONS – CAPITALS
Germinal zone - Helvetica bold
Transient structure - Times bold italic
Permanent structure - Times Roman or Bold

PLATE 2A
CR 32 mm, GW 9.6, C609
Horizontal Section 139

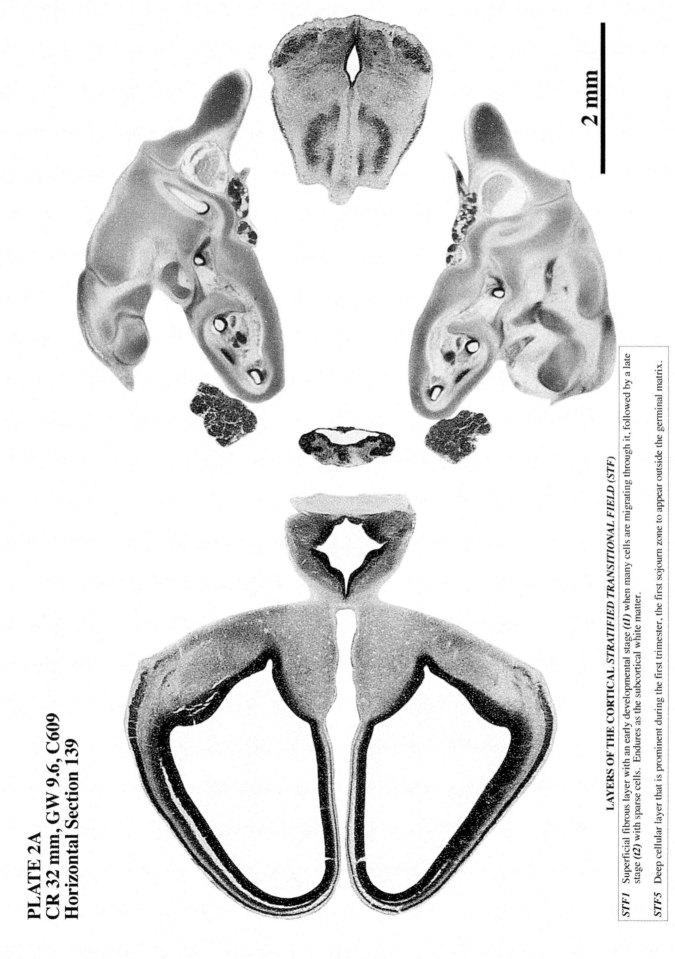

2 mm

LAYERS OF THE CORTICAL *STRATIFIED TRANSITIONAL FIELD* (STF)

STF1 Superficial fibrous layer with an early developmental stage (*t1*) when many cells are migrating through it, followed by a late stage (*t2*) with sparse cells. Endures as the subcortical white matter.

STF5 Deep cellular layer that is prominent during the first trimester, the first sojourn zone to appear outside the germinal matrix.

PLATE 2B

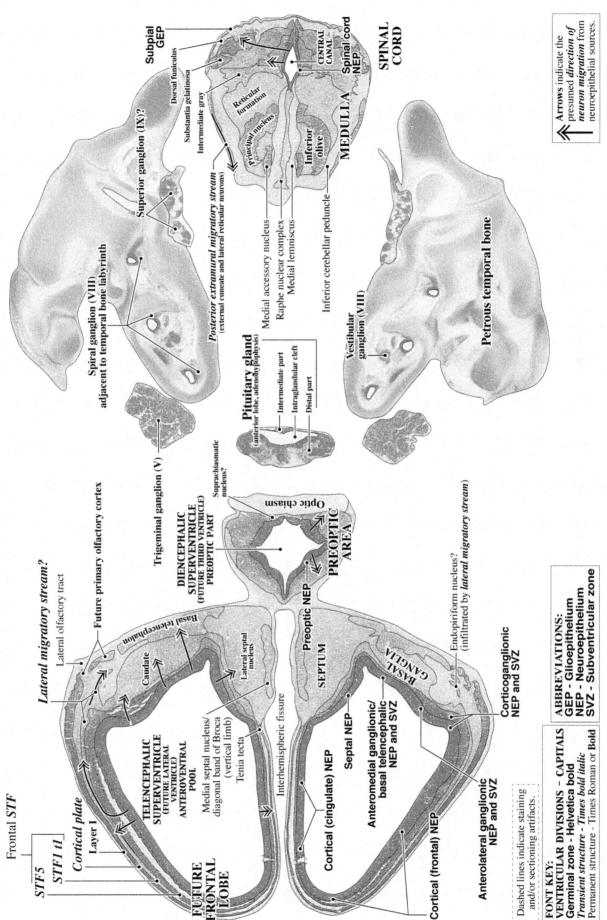

Frontal *STF*

STF5
STF1 t1

Cortical plate
Layer I

Lateral migratory stream?
Lateral olfactory tract
Future primary olfactory cortex

Superior ganglion (IX)?

Subpial GEP

Dorsal funiculus

Substantia gelatinosa
Intermediate gray

Spiral ganglion (VIII)
adjacent to temporal bone labyrinth

Superior ganglion (VIII)

Reticular
formation

Principal nucleus

Posterior extramural migratory stream
(external cuneate and lateral reticular neurons)

CENTRAL
CANAL

Spinal cord
NEP

SPINAL
CORD

Inferior
olive

MEDULLA

Medial accessory nucleus
Raphe nuclear complex
Medial lemniscus

Inferior cerebellar peduncle

Vestibular
ganglion (VIII)

Petrous temporal bone

Trigeminal ganglion (V)

Pituitary gland
(anterior lobe, adenohypophysis)

Intermediate part
Intraglandular cleft
Distal part

Suprachiasmatic
nucleus?

DIENCEPHALIC
SUPERVENTRICLE
(FUTURE THIRD VENTRICLE)
PREOPTIC PART

Optic chiasm

Preoptic NEP

PREOPTIC
AREA

FUTURE
FRONTAL
LOBE

TELENCEPHALIC
SUPERVENTRICLE
(FUTURE LATERAL
VENTRICLE)
ANTEROVENTRAL
POOL

Caudate

Basal telencephalon

Medial septal nucleus/
diagonal band of Broca
(vertical limb)

Tenia tecta

Lateral septal
nucleus

Interhemispheric fissure

SEPTUM

*BASAL
GANGLIA*

Endopiriform nucleus?
(infiltrated by *lateral migratory stream*)

Corticoganglionic
NEP and SVZ

Cortical (cingulate) NEP

Septal NEP

Anteromedial ganglionic/
basal telencephalic
NEP and SVZ

Cortical (frontal) NEP

Anterolateral ganglionic
NEP and SVZ

ABBREVIATIONS:
GEP - Glioepithelium
NEP - Neuroepithelium
SVZ - Subventricular zone

Arrows indicate the
presumed *direction of
neuron migration* from
neuroepithelial sources.

FONT KEY:
VENTRICULAR DIVISIONS - CAPITALS
Germinal zone - Helvetica bold
Transient structure - Times bold italic
Permanent structure - Times Roman or Bold

Dashed lines indicate staining
and/or sectioning artifacts.

23

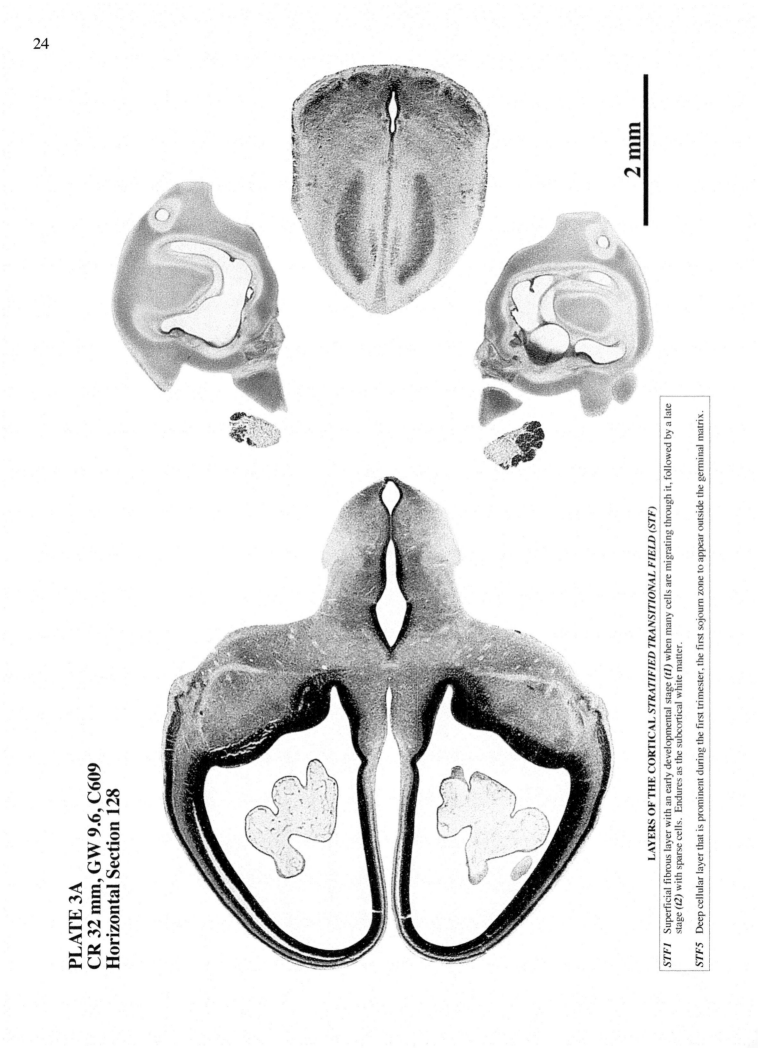

PLATE 3A
CR 32 mm, GW 9.6, C609
Horizontal Section 128

LAYERS OF THE CORTICAL *STRATIFIED TRANSITIONAL FIELD* (STF)

STF1 Superficial fibrous layer with an early developmental stage (*t1*) when many cells are migrating through it, followed by a late stage (*t2*) with sparse cells. Endures as the subcortical white matter.

STF5 Deep cellular layer that is prominent during the first trimester, the first sojourn zone to appear outside the germinal matrix.

2 mm

PLATE 3B

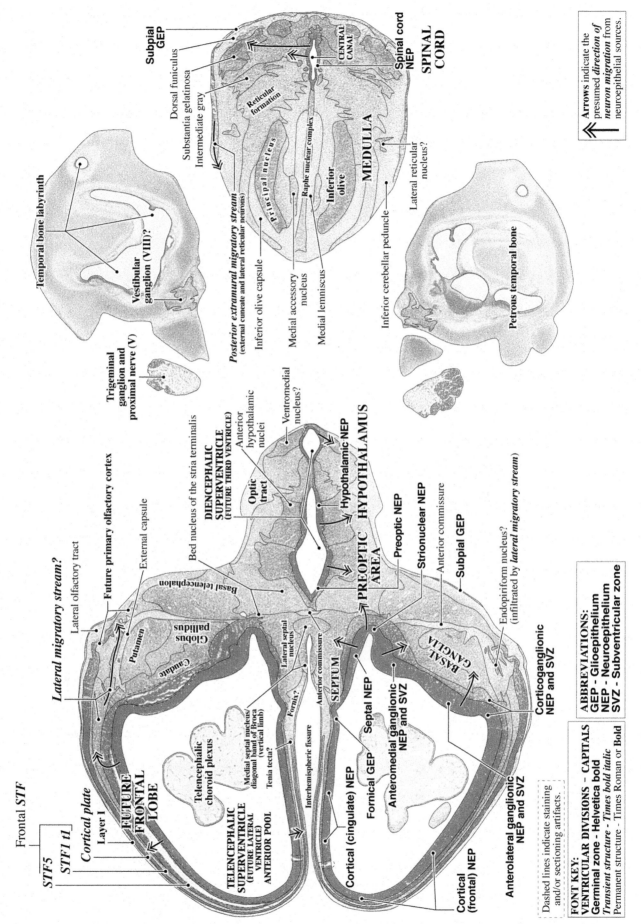

STF5

STF1 t1

Frontal *STF*

Cortical plate

Layer I

FUTURE FRONTAL LOBE

Lateral migratory stream?

Lateral olfactory tract

Future primary olfactory cortex

External capsule

Bed nucleus of the stria terminalis

Patamen

Basal telencephalon

Globus pallidus

Caudate

TELENCEPHALIC SUPERVENTRICLE (FUTURE LATERAL VENTRICLE) ANTERIOR POOL

Telencephalic choroid plexus

Medial septal nucleus/ diagonal band of Broca (vertical limb)

Tenia tecta?

Lateral septal nucleus

Fornix?

Anterior commissure

Interhemispheric fissure

Cortical (cingulate) NEP

Fornical GEP

Septal NEP

Anteromedial ganglionic NEP and SVZ

SEPTUM

Cortical (frontal) NEP

Anterolateral ganglionic NEP and SVZ

Corticoganglionic NEP and SVZ

BASAL GANGLIA

Endopiriform nucleus? (infiltrated by *lateral migratory stream*)

Subpial GEP

Anterior commissure

Strionuclear NEP

Preoptic NEP

PREOPTIC AREA

HYPOTHALAMUS

Hypothalamic NEP

Ventromedial nucleus?

DIENCEPHALIC SUPERVENTRICLE (FUTURE THIRD VENTRICLE)

Anterior hypothalamic nuclei

Optic tract

Temporal bone labyrinth

Vestibular ganglion (VIII)?

Trigeminal ganglion and proximal nerve (V)

Subpial GEP

Dorsal funiculus

Substantia gelatinosa

Intermediate gray

Reticular formation

CENTRAL CANAL

Spinal cord NEP

SPINAL CORD

Posterior extramural migratory stream (external cuneate and lateral reticular neurons)

Inferior olive capsule

Principal nucleus

Raphe nuclear complex

Medial accessory nucleus

Medial lemniscus

Inferior olive

MEDULLA

Inferior cerebellar peduncle

Lateral reticular nucleus?

Petrous temporal bone

Arrows indicate the presumed *direction of neuron migration* from neuroepithelial sources.

Dashed lines indicate staining and/or sectioning artifacts.

FONT KEY:
VENTRICULAR DIVISIONS – CAPITALS
Germinal zone - Helvetica bold
Transient structure - Times bold italic
Permanent structure - Times Roman or Bold

ABBREVIATIONS:
GEP - Glioepithelium
NEP - Neuroepithelium
SVZ - Subventricular zone

25

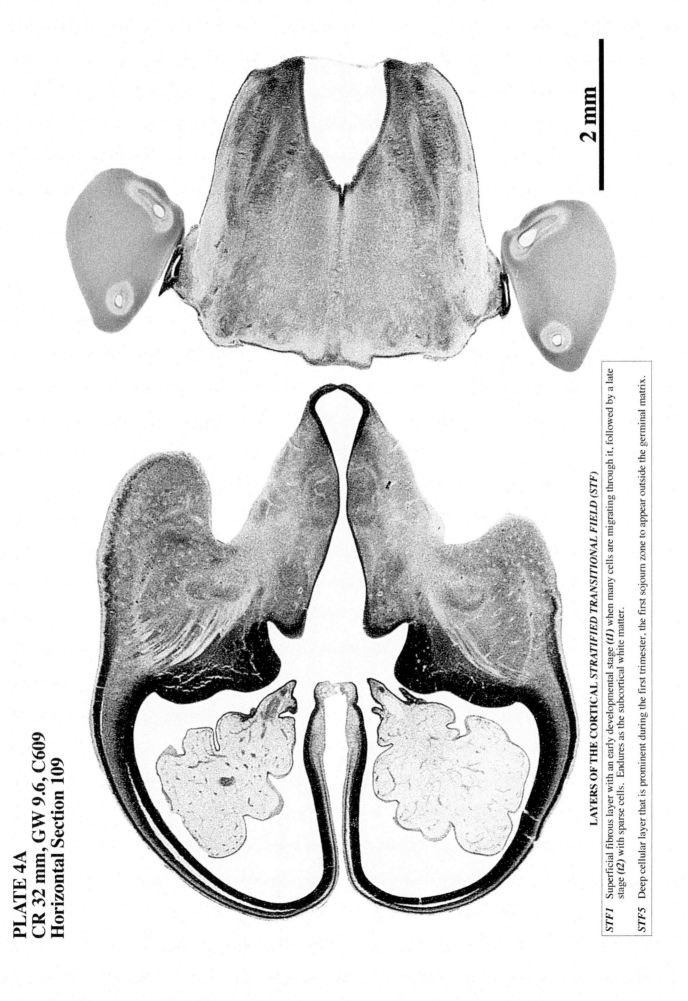

26

PLATE 4A
CR 32 mm, GW 9.6, C609
Horizontal Section 109

2 mm

LAYERS OF THE CORTICAL STRATIFIED TRANSITIONAL FIELD (STF)

STF1 Superficial fibrous layer with an early developmental stage (*t1*) when many cells are migrating through it, followed by a late stage (*t2*) with sparse cells. Endures as the subcortical white matter.

STF5 Deep cellular layer that is prominent during the first trimester, the first sojourn zone to appear outside the germinal matrix.

PLATE 4B

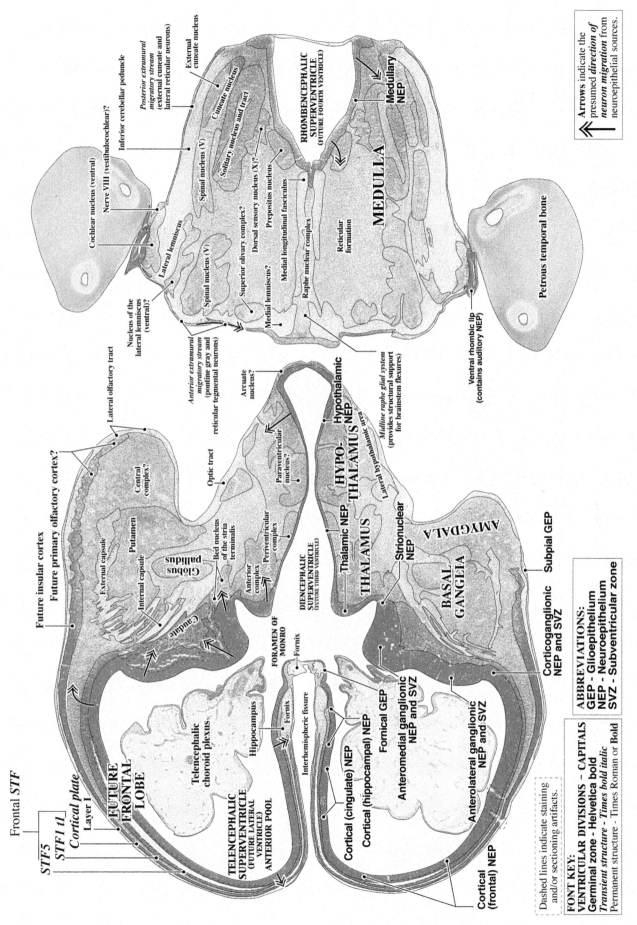

Frontal *STF*

STF5
STF1 tL
Cortical plate
Layer I

FUTURE FRONTAL LOBE

Future insular cortex
Future primary olfactory cortex?

Lateral olfactory tract

External capsule
Internal capsule
Central complex?
Putamen
Globus pallidus
Caudate

Optic tract
Bed nucleus of the stria terminalis
Anterior complex
Periventricular complex
Paraventricular nucleus?
Arcuate nucleus?

Hypothalamic

HYPO-THALAMUS
Lateral hypothalamic area
Thalamic NEP
THALAMUS
Strionuclear NEP

DIENCEPHALIC SUPERVENTRICLE (FUTURE THIRD VENTRICLE)

AMYGDALA
BASAL GANGLIA

Corticoganglionic NEP and SVZ
Subpial GEP

TELENCEPHALIC SUPERVENTRICLE (FUTURE LATERAL VENTRICLE) ANTERIOR POOL

Telencephalic choroid plexus
Hippocampus
Fornix
Fornix
FORAMEN OF MONRO
Interhemispheric fissure

Fornical GEP
Cortical (cingulate) NEP
Cortical (hippocampal) NEP
Anteromedial ganglionic NEP and SVZ
Anterolateral ganglionic NEP and SVZ
Cortical (frontal) NEP

Posterior extramural migratory stream (external cuneate and lateral reticular neurons)
External cuneate nucleus
Cuneate nucleus
Inferior cerebellar peduncle
Nerve VIII (vestibulocochlear)?
Cochlear nucleus (ventral)

RHOMBENCEPHALIC SUPERVENTRICLE (FUTURE FOURTH VENTRICLE)
Medullary NEP
MEDULLA

Spinal nucleus (V)
Solitary nucleus and tract
Superior olivary complex?
Dorsal sensory nucleus (X)?
Prepositus nucleus
Medial longitudinal fasciculus
Raphe nuclear complex
Reticular formation

Spinal nucleus (V)
Lateral lemniscus
Nucleus of the lateral lemniscus (ventral)?
Medial lemniscus?
Medial lemniscus?
Anterior extramural migratory stream (pontine gray and reticular tegmental neurons)

Midline raphe glial system (provides structural support for brainstem flexures)

Ventral rhombic lip (contains auditory NEP)
Petrous temporal bone

Dashed lines indicate staining and/or sectioning artifacts.

FONT KEY:
VENTRICULAR DIVISIONS – CAPITALS
Germinal zone - Helvetica bold
Transient structure - Times bold italic
Permanent structure - Times Roman or Bold

ABBREVIATIONS:
GEP - Glioepithelium
NEP - Neuroepithelium
SVZ - Subventricular zone

Arrows indicate the presumed *direction of neuron migration* from neuroepithelial sources.

27

28

PLATE 5A
CR 32 mm, GW 9.6, C609
Horizontal Section 97

See a high-magnification view of the
diencephalon and basal ganglia
in Plates 11A and B.

2 mm

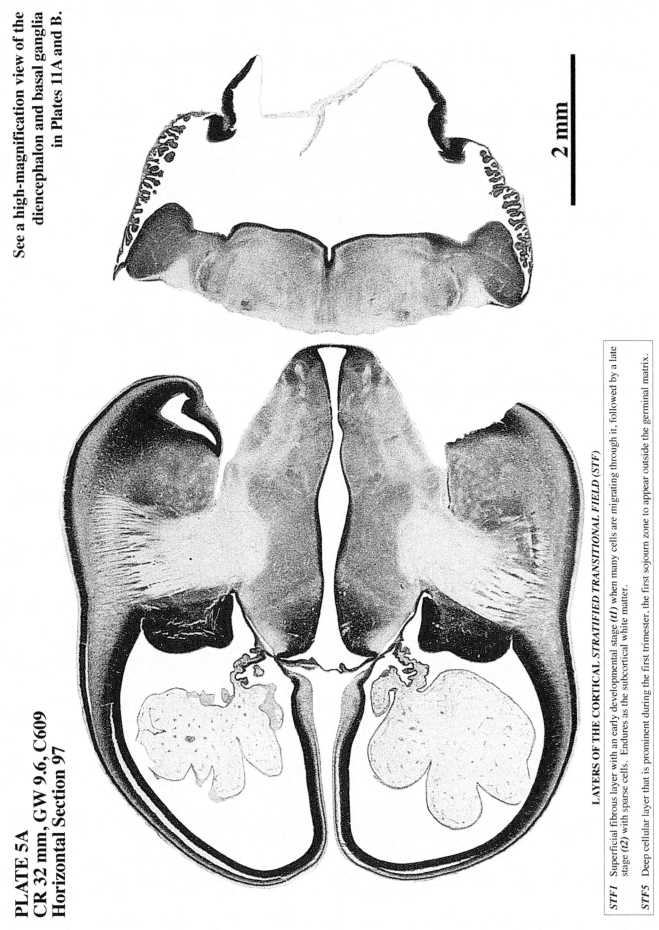

LAYERS OF THE CORTICAL STRATIFIED TRANSITIONAL FIELD (STF)

STF1   Superficial fibrous layer with an early developmental stage (*t1*) when many cells are migrating through it, followed by a late
stage (*t2*) with sparse cells. Endures as the subcortical white matter.

STF5   Deep cellular layer that is prominent during the first trimester, the first sojourn zone to appear outside the germinal matrix.

PLATE 5B

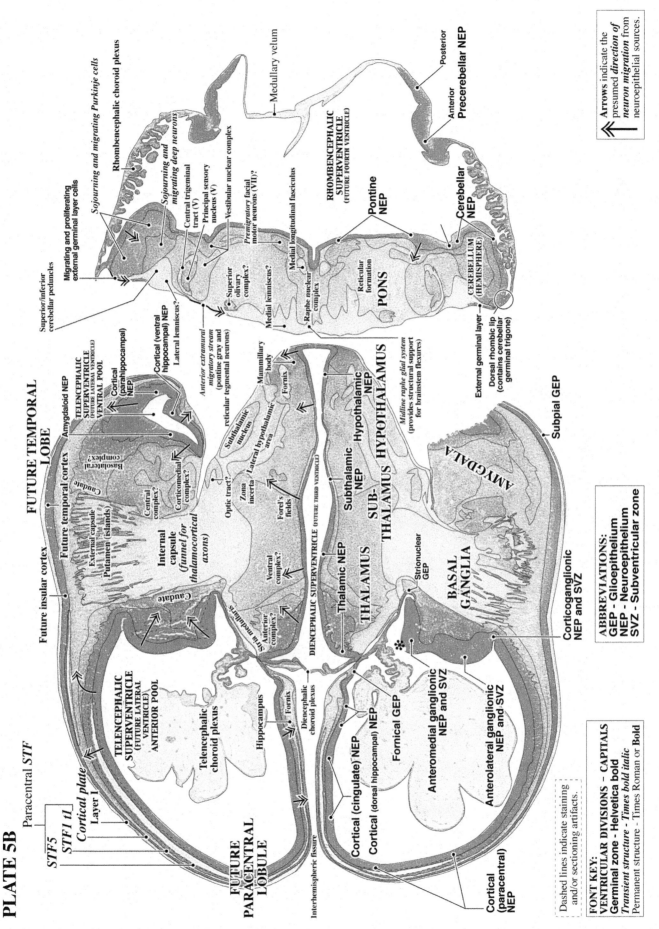

29

Arrows indicate the presumed *direction of neuron migration* from neuroepithelial sources.

Dashed lines indicate staining and/or sectioning artifacts.

FONT KEY:
VENTRICULAR DIVISIONS – CAPITALS
Germinal zone - Helvetica bold
Transient structure - Times bold italic
Permanent structure - Times Roman or Bold

ABBREVIATIONS:
GEP - Glioepithelium
NEP - Neuroepithelium
SVZ - Subventricular zone

30

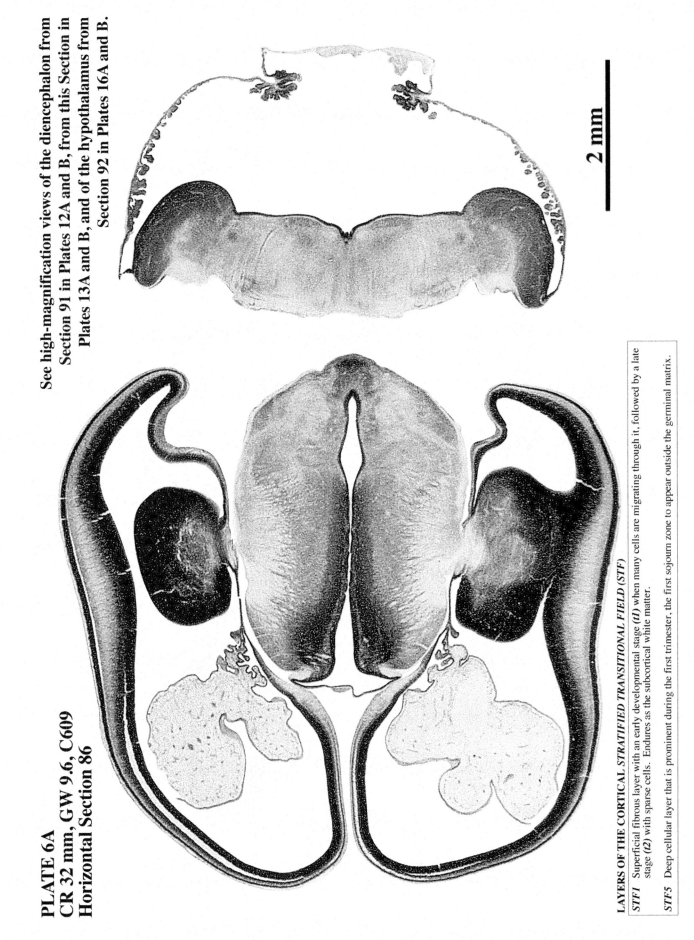

See high-magnification views of the diencephalon from
Section 91 in Plates 12A and B, from this Section in
Plates 13A and B, and of the hypothalamus from
Section 92 in Plates 16A and B.

2 mm

PLATE 6A
CR 32 mm, GW 9.6, C609
Horizontal Section 86

LAYERS OF THE CORTICAL *STRATIFIED TRANSITIONAL FIELD (STF)*

STF1 Superficial fibrous layer with an early developmental stage (*t1*) when many cells are migrating through it, followed by a late
stage (*t2*) with sparse cells. Endures as the subcortical white matter.

STF5 Deep cellular layer that is prominent during the first trimester, the first sojourn zone to appear outside the germinal matrix.

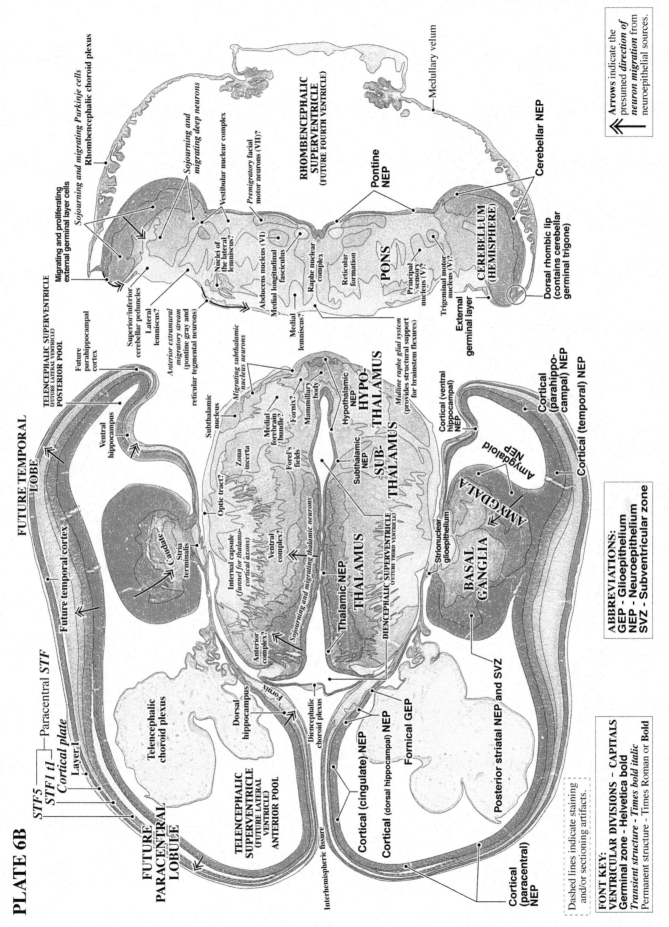

PLATE 6B

31

Arrows indicate the presumed *direction of neuron migration* from neuroepithelial sources.

Dashed lines indicate staining and/or sectioning artifacts.

FONT KEY:
VENTRICULAR DIVISIONS – CAPITALS
Germinal zone – Helvetica bold
Transient structure – *Times bold italic*
Permanent structure – Times Roman or Bold

ABBREVIATIONS:
GEP - Glioepithelium
NEP - Neuroepithelium
SVZ - Subventricular zone

FUTURE TEMPORAL LOBE

Migrating and proliferating external germinal layer cells

Sojourning and migrating Purkinje cells
Rhombencephalic choroid plexus

Sojourning and migrating deep neurons

Vestibular nuclear complex

Premigratory facial motor neurons (VII)?

RHOMBENCEPHALIC SUPERVENTRICLE
(FUTURE FOURTH VENTRICLE)

Pontine NEP

Nuclei of the lateral lemniscus?

Abducens nucleus (VI)

Medial longitudinal fasciculus

Raphe nuclear complex

Reticular formation

PONS

Principal sensory nucleus (V)?

Cerebellar NEP

CEREBELLUM (HEMISPHERE)

Trigeminal motor nucleus (V)?

External germinal layer

Dorsal rhombic lip (contains cerebellar germinal trigone)

Medullary velum

TELENCEPHALIC SUPERVENTRICLE
(FUTURE LATERAL VENTRICLE)
POSTERIOR POOL

Future parahippocampal cortex

Ventral hippocampus

Superior/inferior cerebellar peduncles

Lateral lemniscus?

Anterior extramural migratory stream (pontine gray and reticular tegmental neurons)

Medial lemniscus?

Migrating subthalamic nucleus neurons

Subthalamic nucleus

Medial forebrain bundle?

Fornix?

Mammillary body

Hypothalamic NEP

HYPO-THALAMUS

Subthalamic NEP

SUB-THALAMUS

Midline raphe glial system (provides structural support for brainstem flexures)

FUTURE TEMPORAL LOBE

Ventral hippocampus

Optic tract?

Zona incerta

Forel's fields

Internal capsule (funnel for thalamo-cortical axons)

Ventral complex?

Sojourning and migrating thalamic neurons

Thalamic NEP

THALAMUS

DIENCEPHALIC SUPERVENTRICLE
(FUTURE THIRD VENTRICLE)

Cortical (ventral hippocampal) NEP

Cortical (parahippo-campal) NEP

Amygdaloid NEP

Cortical (temporal) NEP

Future temporal cortex

Caudate

Stria terminalis

Anterior complex?

AMYGDALA

Strionuclear glioepithelium

BASAL GANGLIA

STF5
STF1 t1 — Paracentral STF
Cortical plate
Layer 1

FUTURE PARACENTRAL LOBULE

Telencephalic choroid plexus

Dorsal hippocampus

Fornix

Diencephalic choroid plexus

TELENCEPHALIC SUPERVENTRICLE
(FUTURE LATERAL VENTRICLE)
ANTERIOR POOL

Interhemispheric fissure

Cortical (cingulate) NEP

Cortical (dorsal hippocampal) NEP

Fornical GEP

Posterior striatal NEP and SVZ

Cortical (paracentral) NEP

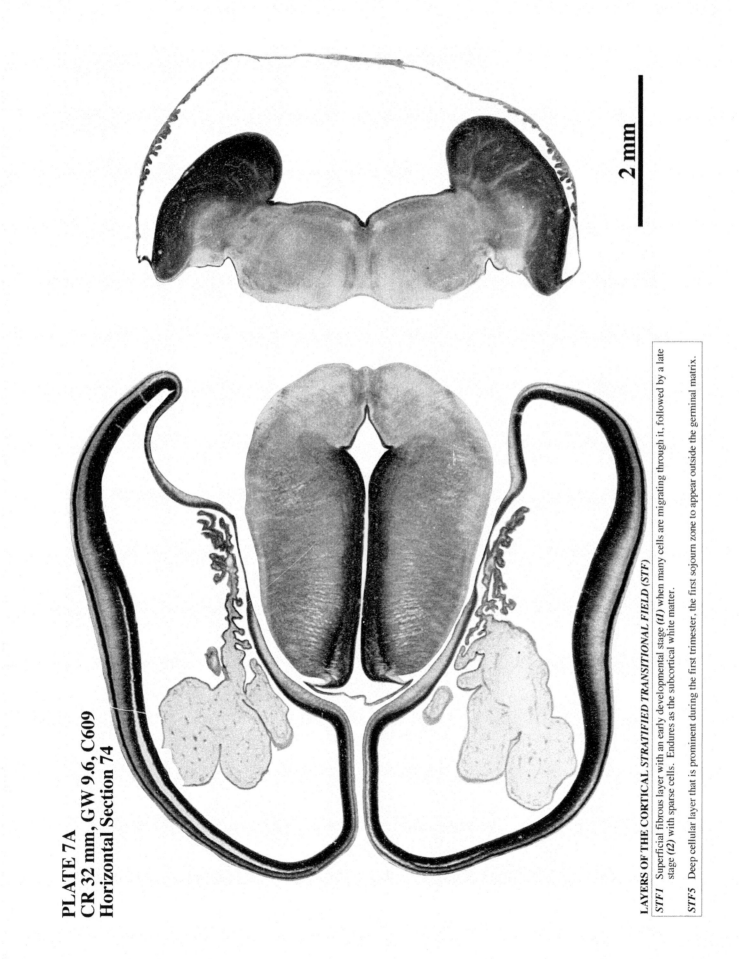

32

PLATE 7A
CR 32 mm, GW 9.6, C609
Horizontal Section 74

2 mm

LAYERS OF THE CORTICAL STRATIFIED TRANSITIONAL FIELD (STF)

STF1 Superficial fibrous layer with an early developmental stage (*t1*) when many cells are migrating through it, followed by a late stage (*t2*) with sparse cells. Endures as the subcortical white matter.

STF5 Deep cellular layer that is prominent during the first trimester, the first sojourn zone to appear outside the germinal matrix.

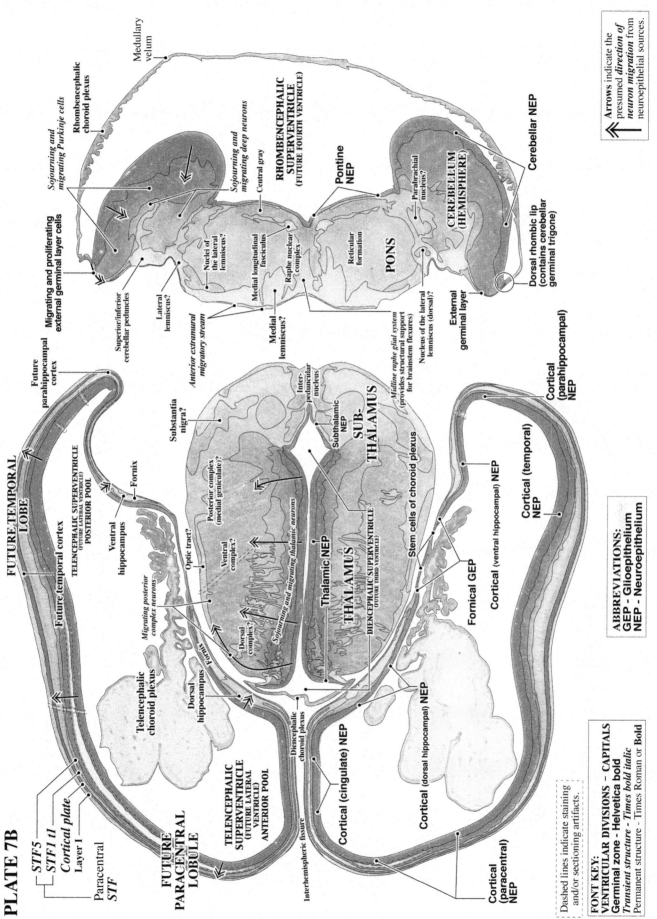

PLATE 7B

33

Arrows indicate the presumed *direction of neuron migration* from neuroepithelial sources.

ABBREVIATIONS:
GEP - Glioepithelium
NEP - Neuroepithelium

FONT KEY:
VENTRICULAR DIVISIONS – CAPITALS
Germinal zone - Helvetica bold
Transient structure - Times bold italic
Permanent structure - Times Roman or Bold

Dashed lines indicate staining and/or sectioning artifacts.

Medullary velum

Rhombencephalic choroid plexus

Sojourning and migrating Purkinje cells

Sojourning and migrating deep neurons

Central gray

RHOMBENCEPHALIC SUPERVENTRICLE (FUTURE FOURTH VENTRICLE)

Pontine NEP

Cerebellar NEP

Parabrachial nucleus?

CEREBELLUM (HEMISPHERE)

Migrating and proliferating external germinal layer cells

Superior/inferior cerebellar peduncles

Lateral lemniscus?

Nuclei of the lateral lemniscus?

Medial longitudinal fasciculus

Raphe nuclear complex

Reticular formation

PONS

Nuclei of the lateral lemniscus (dorsal)?

Dorsal rhombic lip (contains cerebellar germinal trigone)

External germinal layer

Anterior extramural migratory stream

Medial lemniscus?

Midline raphe glial system (provides structural support for brainstem flexures)

Future parahippocampal cortex

FUTURE TEMPORAL LOBE

Future temporal cortex

TELENCEPHALIC SUPERVENTRICLE (FUTURE LATERAL VENTRICLE) POSTERIOR POOL

Fornix

Ventral hippocampus

Substantia nigra?

Inter-peduncular nucleus

Subthalamic NEP

Posterior complex (medial geniculate)?

SUB-THALAMUS

Cortical (parahippocampal) NEP

STF5
STF1 t1
Cortical plate
Layer I
Paracentral *STF*

Telencephalic choroid plexus

Migrating posterior complex neurons

Optic tract?

Ventral complex?

Dorsal complex?

Sojourning and migrating thalamic neurons

Thalamic NEP

THALAMUS

Stem cells of choroid plexus

Cortical (temporal) NEP

Dorsal hippocampus

Fornix

DIENCEPHALIC SUPERVENTRICLE (FUTURE THIRD VENTRICLE)

Fornical GEP

Cortical (ventral hippocampal) NEP

FUTURE PARACENTRAL LOBULE

TELENCEPHALIC SUPERVENTRICLE (FUTURE LATERAL VENTRICLE) ANTERIOR POOL

Diencephalic choroid plexus

Cortical (cingulate) NEP

Cortical (dorsal hippocampal) NEP

Interhemispheric fissure

Cortical (paracentral) NEP

34

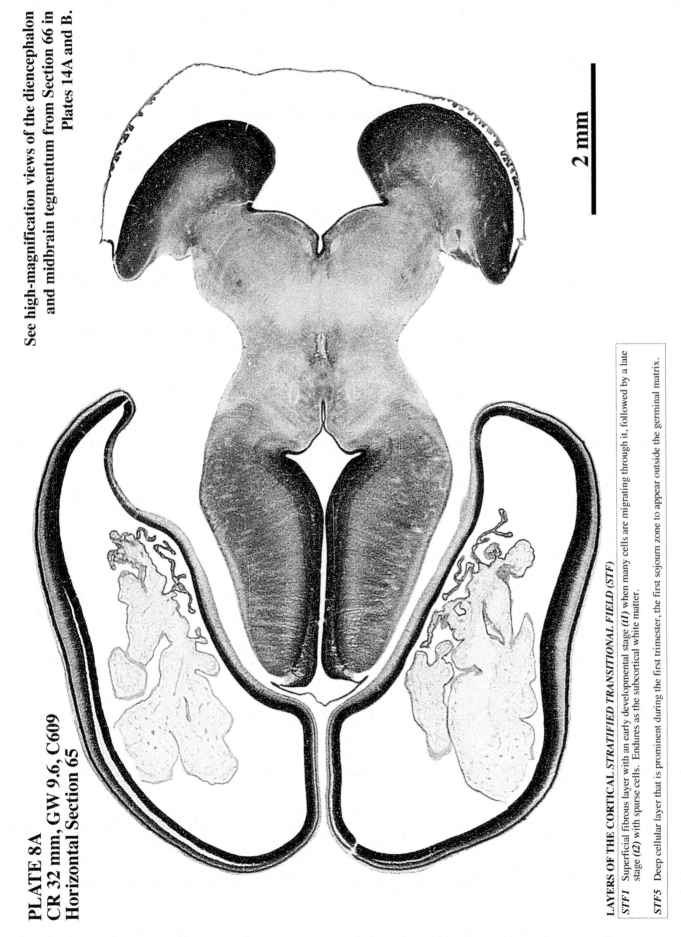

PLATE 8A
CR 32 mm, GW 9.6, C609
Horizontal Section 65

See high-magnification views of the diencephalon
and midbrain tegmentum from Section 66 in
Plates 14A and B.

2 mm

LAYERS OF THE CORTICAL *STRATIFIED TRANSITIONAL FIELD* (STF)

STF1 Superficial fibrous layer with an early developmental stage (*t1*) when many cells are migrating through it, followed by a late
stage (*t2*) with sparse cells. Endures as the subcortical white matter.

STF5 Deep cellular layer that is prominent during the first trimester, the first sojourn zone to appear outside the germinal matrix.

PLATE 8B

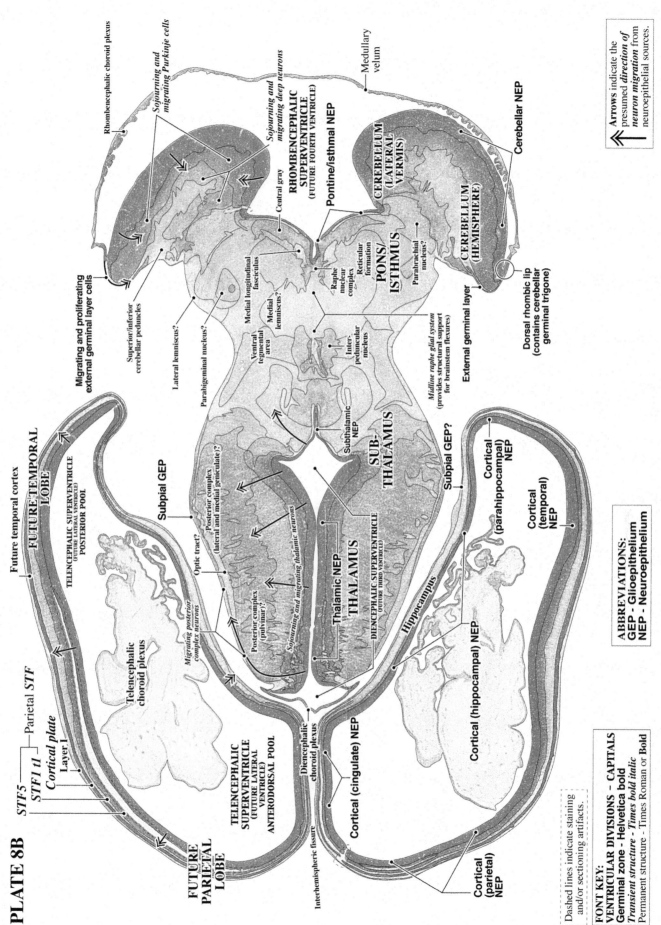

STF5 ── STF

STF11 t1

Cortical plate

Layer I

Parietal STF

Future temporal cortex

FUTURE TEMPORAL LOBE

Telencephalic choroid plexus

TELENCEPHALIC SUPERVENTRICLE (FUTURE LATERAL VENTRICLE) POSTERIOR POOL

Subpial GEP

Migrating posterior complex neurons

FUTURE PARIETAL LOBE

Cortical (parietal) NEP

TELENCEPHALIC SUPERVENTRICLE (FUTURE LATERAL VENTRICLE) ANTERODORSAL POOL

Diencephalic choroid plexus

Interhemispheric fissure

Cortical (cingulate) NEP

Optic tract?

Posterior complex (lateral and medial geniculate)?

Posterior complex (pulvinar)?

Sojourning and migrating thalamic neurons

Thalamic NEP

THALAMUS

DIENCEPHALIC SUPERVENTRICLE (FUTURE THIRD VENTRICLE)

Subthalamic NEP

SUB-THALAMUS

Subpial GEP?

Hippocampus

Cortical (hippocampal) NEP

Cortical (parahippocampal) NEP

Cortical (temporal) NEP

Rhombencephalic choroid plexus

Sojourning and migrating Purkinje cells

Migrating and proliferating external germinal layer cells

Superior/inferior cerebellar peduncles

Lateral lemniscus?

Parabigeminal nucleus?

Ventral tegmental area

Medial longitudinal fasciculus

Medial lemniscus?

Central gray

Sojourning and migrating deep neurons

RHOMBENCEPHALIC SUPERVENTRICLE (FUTURE FOURTH VENTRICLE)

Pontine/isthmal NEP

Raphe nuclear complex

Reticular formation

Inter-peduncular nucleus

PONS/ ISTHMUS

CEREBELLUM (LATERAL VERMIS)

CEREBELLUM (HEMISPHERE)

Parabrachial nucleus?

Midline raphe glial system (provides structural support for brainstem flexures)

External germinal layer

Dorsal rhombic lip (contains cerebellar germinal trigone)

Cerebellar NEP

Medullary velum

Arrows indicate the presumed *direction of neuron migration* from neuroepithelial sources.

Dashed lines indicate staining and/or sectioning artifacts.

FONT KEY:
VENTRICULAR DIVISIONS – CAPITALS
Germinal zone - Helvetica bold
Transient structure - Times bold italic
Permanent structure - Times Roman or Bold

ABBREVIATIONS:
GEP - Glioepithelium
NEP - Neuroepithelium

PLATE 9A
CR 32 mm, GW 9.6, C609
Horizontal Section 45

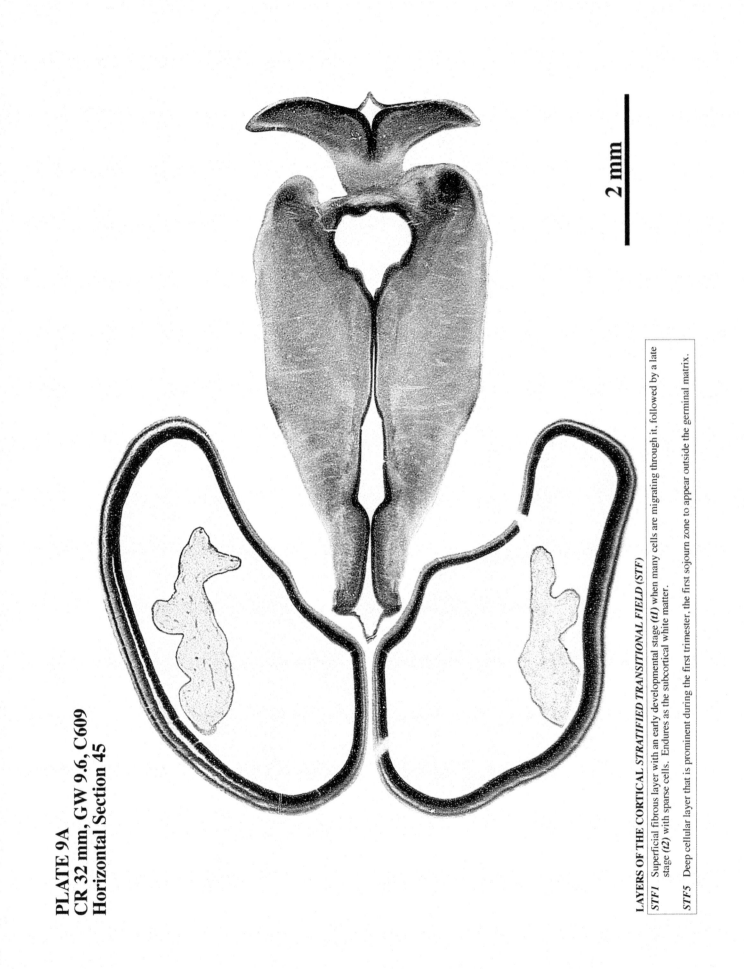

2 mm

LAYERS OF THE CORTICAL *STRATIFIED TRANSITIONAL FIELD (STF)*

STF1 Superficial fibrous layer with an early developmental stage (*t1*) when many cells are migrating through it, followed by a late stage (*t2*) with sparse cells. Endures as the subcortical white matter.

STF5 Deep cellular layer that is prominent during the first trimester, the first sojourn zone to appear outside the germinal matrix.

PLATE 9B

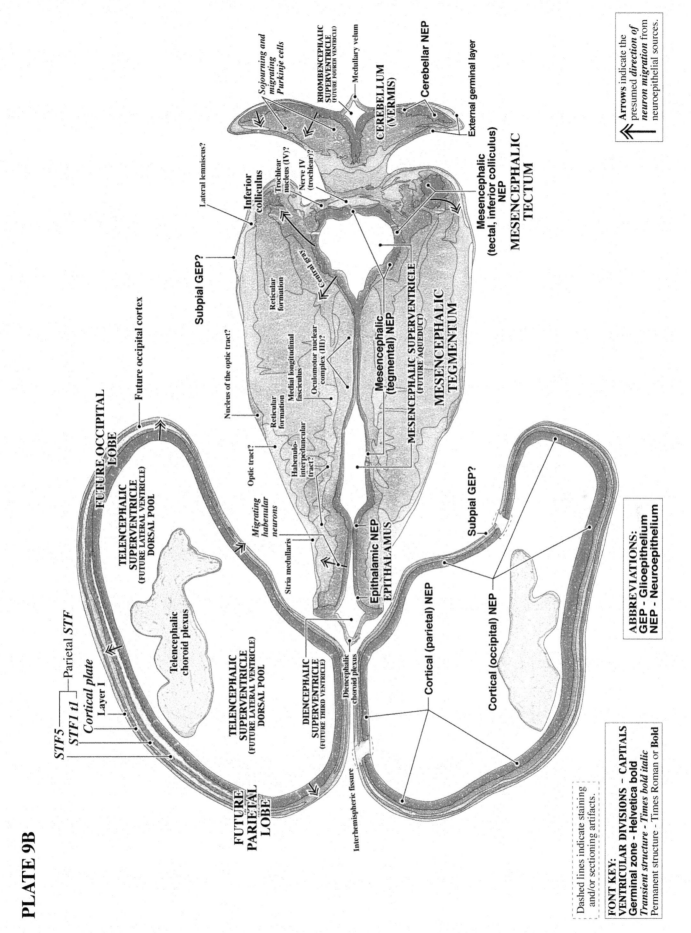

STF5
STF1 t1
Cortical plate
Layer 1
Parietal STF
Cortical plate

FUTURE OCCIPITAL LOBE

Future occipital cortex

TELENCEPHALIC SUPERVENTRICLE
(FUTURE LATERAL VENTRICLE)
DORSAL POOL

Telencephalic choroid plexus

TELENCEPHALIC SUPERVENTRICLE
(FUTURE LATERAL VENTRICLE)
DORSAL POOL

DIENCEPHALIC SUPERVENTRICLE
(FUTURE THIRD VENTRICLE)

Diencephalic choroid plexus

Interhemispheric fissure

FUTURE PARIETAL LOBE

Subpial GEP?

Nucleus of the optic tract?

Lateral lemniscus?

Inferior colliculus

Reticular formation

Cerebral aqueduct

Trochlear nucleus (IV)?

Nerve IV (trochlear)?

Sojourning and migrating Purkinje cells

RHOMBENCEPHALIC SUPERVENTRICLE
(FUTURE FOURTH VENTRICLE)

Medullary velum

CEREBELLUM (VERMIS)

Cerebellar NEP

External germinal layer

Mesencephalic (tectal, inferior colliculus) NEP

MESENCEPHALIC TECTUM

Reticular formation

Medial longitudinal fasciculus

Oculomotor nuclear complex (III)?

Mesencephalic (tegmental) NEP

MESENCEPHALIC SUPERVENTRICLE
(FUTURE AQUEDUCT)

MESENCEPHALIC TEGMENTUM

Optic tract?

Habenulo-interpeduncular tract?

Migrating habenular neurons

Stria medullaris

Epithalamic NEP

EPITHALAMUS

Subpial GEP?

Cortical (parietal) NEP

Cortical (occipital) NEP

Dashed lines indicate staining and/or sectioning artifacts.

FONT KEY:
VENTRICULAR DIVISIONS - CAPITALS
Germinal zone - **Helvetica bold**
Transient structure - Times bold italic
Permanent structure - Times Roman or **Bold**

ABBREVIATIONS:
GEP - Glioepithelium
NEP - Neuroepithelium

Arrows indicate the presumed *direction of neuron migration* from neuroepithelial sources.

38

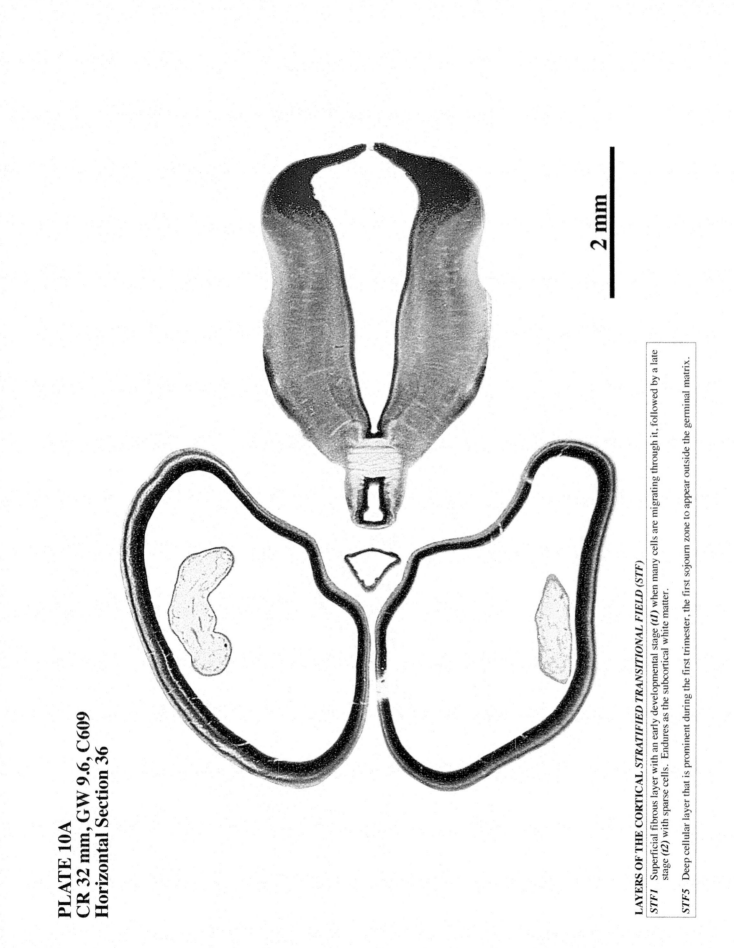

PLATE 10A
CR 32 mm, GW 9.6, C609
Horizontal Section 36

2 mm

LAYERS OF THE CORTICAL *STRATIFIED TRANSITIONAL FIELD* (STF)

STF1   Superficial fibrous layer with an early developmental stage (*t1*) when many cells are migrating through it, followed by a late stage (*t2*) with sparse cells.   Endures as the subcortical white matter.

STF5   Deep cellular layer that is prominent during the first trimester, the first sojourn zone to appear outside the germinal matrix.

PLATE 10B

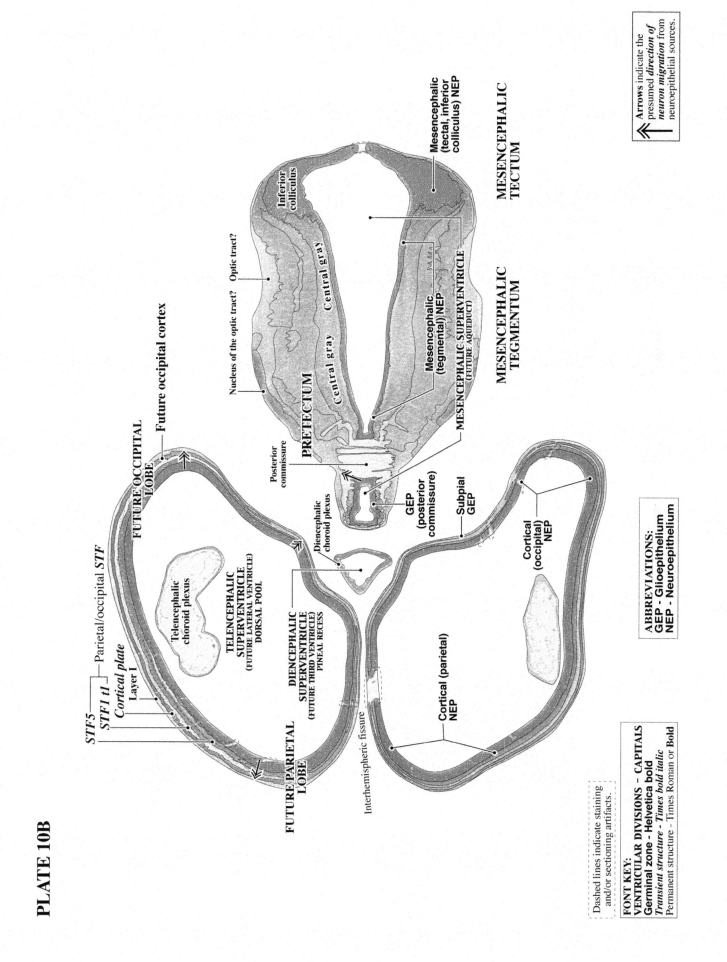

39

Arrows indicate the presumed *direction of neuron migration* from neuroepithelial sources.

Future occipital cortex

FUTURE OCCIPITAL LOBE

Parietal/occipital *STF*

STF5
STF1 t1
Cortical plate
Layer 1

Telencephalic choroid plexus

TELENCEPHALIC SUPERVENTRICLE
(FUTURE LATERAL VENTRICLE)
DORSAL POOL

DIENCEPHALIC SUPERVENTRICLE
(FUTURE THIRD VENTRICLE)
PINEAL RECESS

Diencephalic choroid plexus

Interhemispheric fissure

FUTURE PARIETAL LOBE

Cortical (parietal) NEP

Cortical (occipital) NEP

GEP (posterior commissure)

Subpial GEP

Posterior commissure

PRETECTUM

Optic tract?

Nucleus of the optic tract?

Inferior colliculus

Central gray

Central gray

Mesencephalic (tectal, inferior colliculus) NEP

Mesencephalic (tegmental) NEP

MESENCEPHALIC SUPERVENTRICLE
(FUTURE AQUEDUCT)

MESENCEPHALIC TECTUM

MESENCEPHALIC TEGMENTUM

ABBREVIATIONS:
GEP - Glioepithelium
NEP - Neuroepithelium

Dashed lines indicate staining and/or sectioning artifacts.

FONT KEY:
VENTRICULAR DIVISIONS – CAPITALS
Germinal zone - Helvetica bold
Transient structure - Times bold italic
Permanent structure - Times Roman or **Bold**

PLATE 11A

CR 32 mm, GW 9.6, C609
Horizontal Section 97
**DIENCEPHALON AND
BASAL GANGLIA**

**LAYERS OF THE
CORTICAL STRATIFIED
TRANSITIONAL FIELD (STF)**

STF1 Superficial fibrous layer with an
early developmental stage (*t1*)
when many cells are migrating
through it, followed by a late
stage (*t2*) with sparse cells.
Endures as the subcortical white
matter.

STF5 Deep cellular layer that is
prominent during the first
trimester, the first sojourn zone
to appear outside the germinal
matrix.

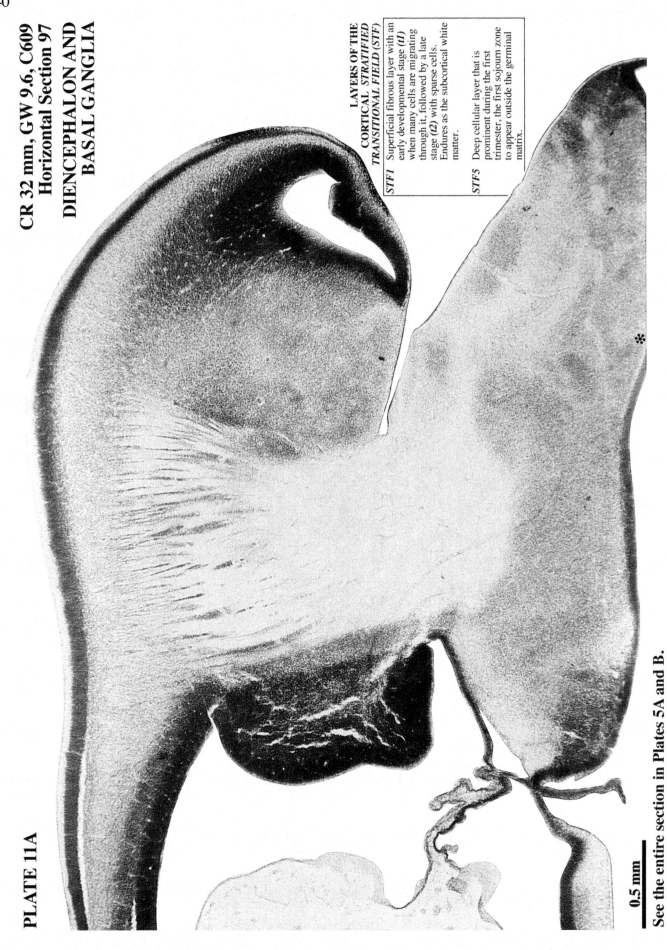

0.5 mm

See the entire section in Plates 5A and B.

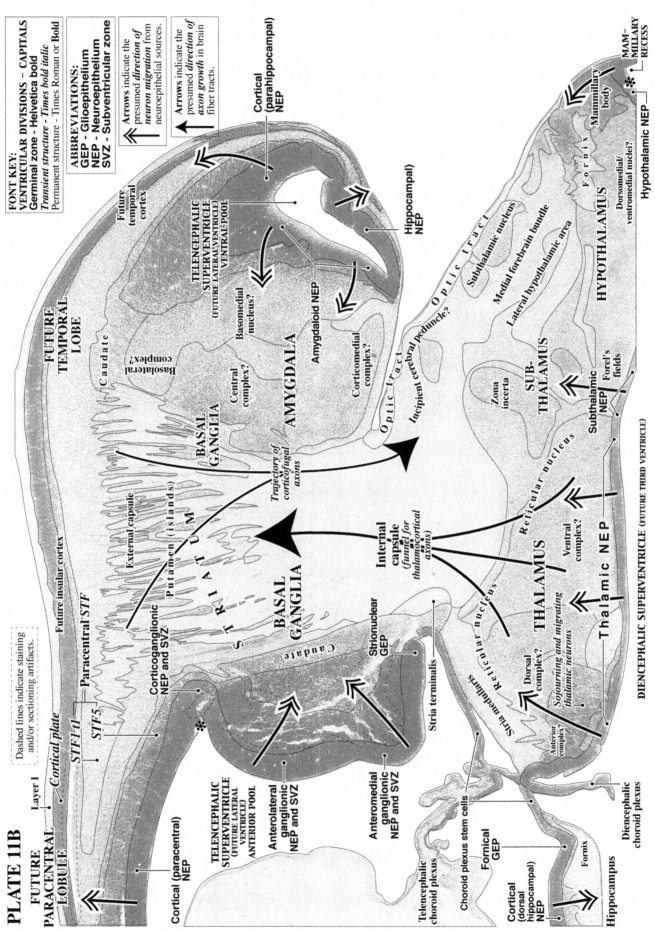

PLATE 11B

FONT KEY:
VENTRICULAR DIVISIONS – CAPITALS
Germinal zone - Helvetica bold
Transient structure - Times bold italic
Permanent structure - Times Roman or Bold

ABBREVIATIONS:
GEP - Glioepithelium
NEP - Neuroepithelium
SVZ - Subventricular zone

Arrows indicate the presumed *direction of neuron migration* from neuroepithelial sources.

Arrows indicate the presumed *direction of axon growth* in brain fiber tracts.

Dashed lines indicate staining and/or sectioning artifacts.

FUTURE PARACENTRAL LOBULE
Layer I
Cortical plate
Future insular cortex
STF-1t1
Paracentral STF
STF-5
Cortical (paracentral) NEP

TELENCEPHALIC SUPERVENTRICLE
(FUTURE LATERAL VENTRICLE ANTERIOR POOL)

Anterolateral ganglionic NEP and SVZ

Anteromedial ganglionic NEP and SVZ

Cortical (parahippocampal) NEP

Future temporal cortex

FUTURE TEMPORAL LOBE

Caudate

Basolateral complex?

TELENCEPHALIC SUPERVENTRICLE
(FUTURE LATERAL VENTRICLE VENTRAL POOL)

Basomedial nucleus?

BASAL GANGLIA

Central complex?

Amygdaloid NEP

AMYGDALA

Corticomedial complex?

Hippocampal NEP

External capsule

BASAL GANGLIA

S T R I A T U M

Putamen (islands)

Corticoganglionic NEP and SVZ

Caudate

Strionuclear GEP

Trajectory of corticofugal axons

Internal capsule (funnel for thalamocortical axons)

O p t i c   t r a c t

Incipient cerebral peduncle?

Subthalamic nucleus

Medial forebrain bundle

Lateral hypothalamic area

Zona incerta

SUB-THALAMUS

Subthalamic NEP
Forel's fields

Mammillary body

F o r n i x

HYPOTHALAMUS

Dorsomedial/ventromedial nuclei?

Hypothalamic NEP

MAM-MILLARY RECESS

Reticular nucleus

Ventral complex?

Thalamic NEP

THALAMUS

Dorsal complex?

Sojourning and migrating thalamic neurons

Dorsal Reticular nucleus

Stria medullaris

Anterior complex?

DIENCEPHALIC SUPERVENTRICLE (FUTURE THIRD VENTRICLE)

Stria terminalis

Telencephalic choroid plexus

Choroid plexus stem cells

Fornical GEP

Fornix

Fornix

Cortical (dorsal hippocampal) NEP

Diencephalic choroid plexus

Hippocampus

41

CR 32 mm, GW 9.6, C609
Horizontal Section 91

DIENCEPHALON

See Plates 5 and 6
for nearby complete sections.

0.5 mm

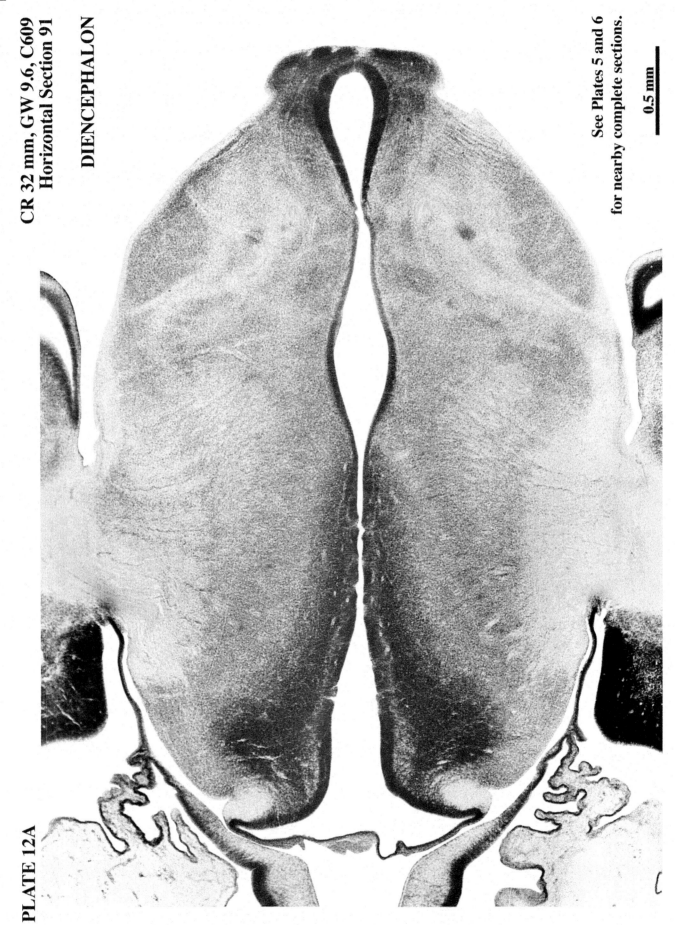

PLATE 12A

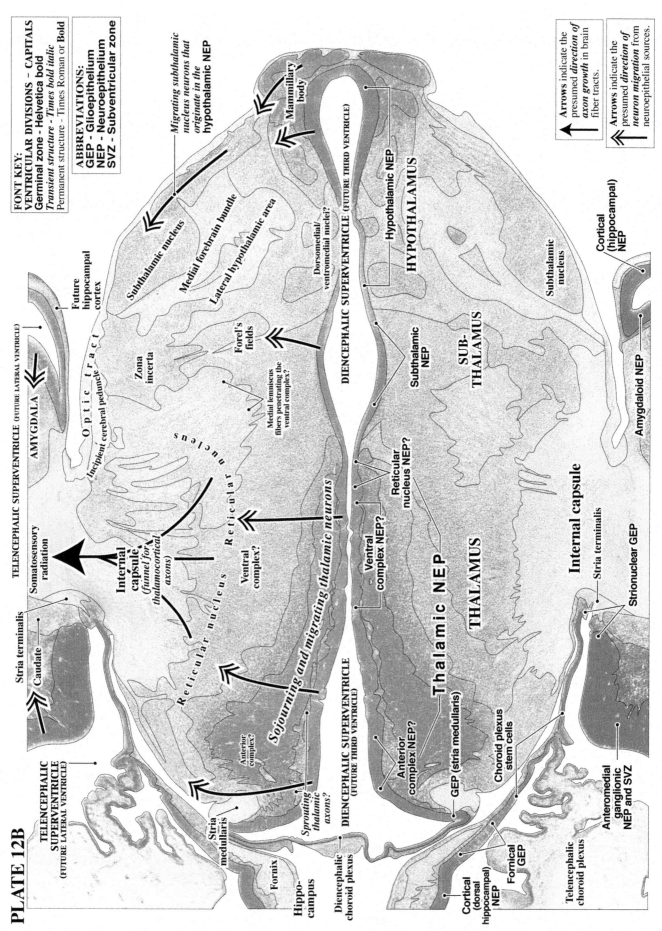

PLATE 12B

43

FONT KEY:
VENTRICULAR DIVISIONS – CAPITALS
Germinal zone – **Helvetica bold**
Transient structure – Times bold italic
Permanent structure – Times Roman or Bold

ABBREVIATIONS:
GEP - Glioepithelium
NEP - Neuroepithelium
SVZ - Subventricular zone

Arrows indicate the presumed *direction of axon growth* in brain fiber tracts.

Arrows indicate the presumed *direction of neuron migration* from neuroepithelial sources.

TELENCEPHALIC SUPERVENTRICLE (FUTURE LATERAL VENTRICLE)

Migrating subthalamic nucleus neurons that originate in the hypothalamic NEP

Subthalamic nucleus

Medial forebrain bundle

Lateral hypothalamic area

Mammillary body

Dorsomedial/ ventromedial nuclei?

DIENCEPHALIC SUPERVENTRICLE (FUTURE THIRD VENTRICLE)

Hypothalamic NEP

HYPOTHALAMUS

Subthalamic NEP

SUB-THALAMUS

Subthalamic nucleus

Cortical (hippocampal) NEP

Future hippocampal cortex

AMYGDALA

O p t i c   t r a c t

Incipient cerebral peduncle

Zona incerta

Forel's fields

Medial lemniscus fibers penetrating the ventral complex?

R e t i c u l a r   n u c l e u s

Ventral complex?

Reticular nucleus NEP?

Ventral complex NEP?

Thalamic NEP

THALAMUS

Amygdaloid NEP

Somatosensory radiation

Internal capsule (funnel for thalamocortical axons)

R e t i c u l a r   n u c l e u s

Sojourning and migrating thalamic neurons

Stria terminalis

Caudate

Internal capsule

Stria terminalis

Strionuclear GEP

TELENCEPHALIC SUPERVENTRICLE (FUTURE LATERAL VENTRICLE)

Anterior complex?

Sprouting thalamic axons?

Stria medullaris

Fornix

Hippocampus

Diencephalic choroid plexus

DIENCEPHALIC SUPERVENTRICLE (FUTURE THIRD VENTRICLE)

Anterior complex NEP?

GEP (stria medullaris)

Choroid plexus stem cells

Cortical (dorsal hippocampal) NEP

Fornical GEP

Telencephalic choroid plexus

Anteromedial ganglionic NEP and SVZ

CR 32 mm, GW 9.6, C609
Horizontal Section 86

DIENCEPHALON

See the entire
Section 86 in
Plates 6A and B.

0.5 mm

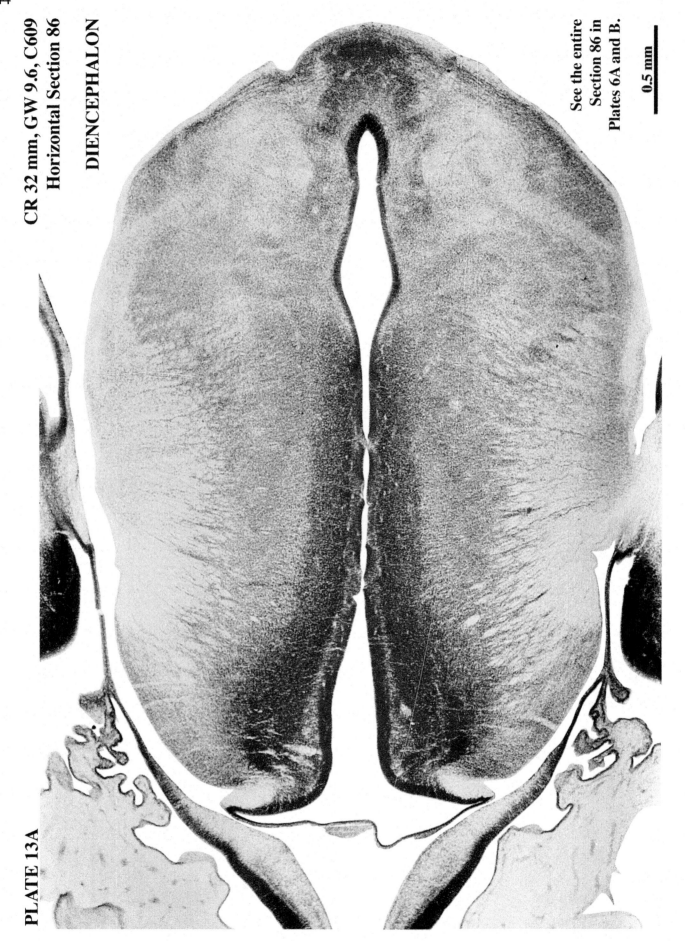

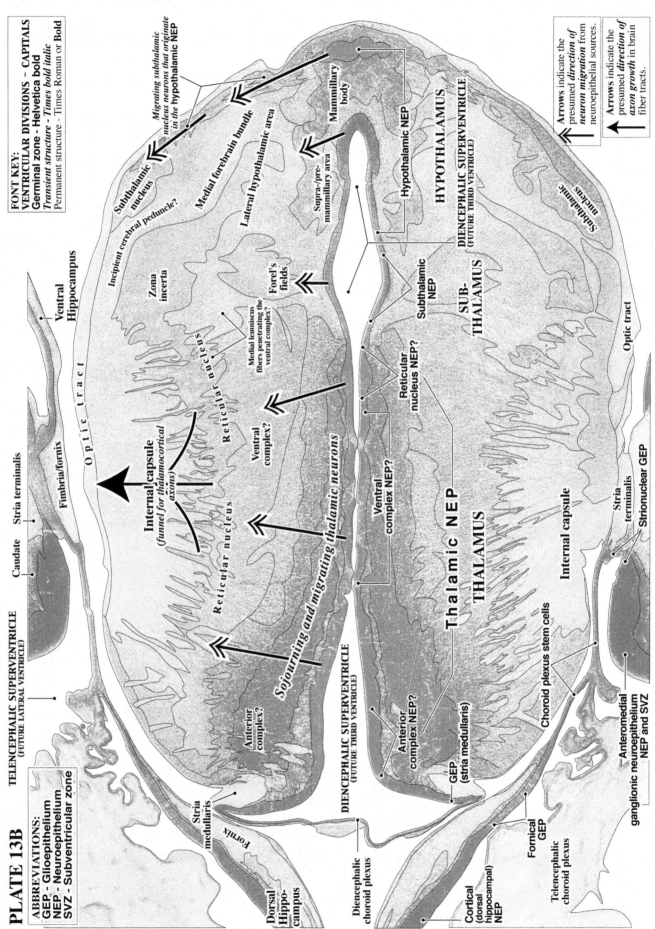

PLATE 13B

TELENCEPHALIC SUPERVENTRICLE
(FUTURE LATERAL VENTRICLE)

ABBREVIATIONS:
GEP - Glioepithelium
NEP - Neuroepithelium
SVZ - Subventricular zone

FONT KEY:
VENTRICULAR DIVISIONS – CAPITALS
Germinal zone - Helvetica bold
Transient structure - Times bold italic
Permanent structure - Times Roman or Bold

Arrows indicate the presumed *direction of neuron migration* from neuroepithelial sources.

Arrows indicate the presumed *direction of axon growth* in brain fiber tracts.

Migrating subthalamic nucleus neurons that originate in the hypothalamic NEP

Mammillary body

Hypothalamic NEP

HYPOTHALAMUS

DIENCEPHALIC SUPERVENTRICLE
(FUTURE THIRD VENTRICLE)

Subthalamic nucleus

Subthalamic nucleus

Medial forebrain bundle

Lateral hypothalamic area

Supra-/pre-mammillary area

Incipient cerebral peduncle?

Zona incerta

Forel's fields

Subthalamic NEP

SUB-THALAMUS

Medial lemniscus fibers penetrating the ventral complex?

Ventral Hippocampus

Optic tract

Reticular nucleus

Reticular nucleus

Reticular nucleus

Reticular nucleus NEP?

Ventral complex?

Sojourning and migrating thalamic neurons

Internal capsule
(funnel for thalamocortical axons)

Ventral complex NEP?

Thalamic NEP

THALAMUS

Optic tract

Fimbria/fornix

Caudate   Stria terminalis

Stria medullaris

Fornix

Dorsal Hippo-campus

Anterior complex?

Anterior complex NEP?

GEP (stria medullaris)

Choroid plexus stem cells

Internal capsule

Stria terminalis

Strionuclear GEP

DIENCEPHALIC SUPERVENTRICLE
(FUTURE THIRD VENTRICLE)

Diencephalic choroid plexus

Cortical (dorsal hippocampal) NEP

Fornical GEP

Telencephalic choroid plexus

Anteromedial ganglionic neuroepithelium NEP and SVZ

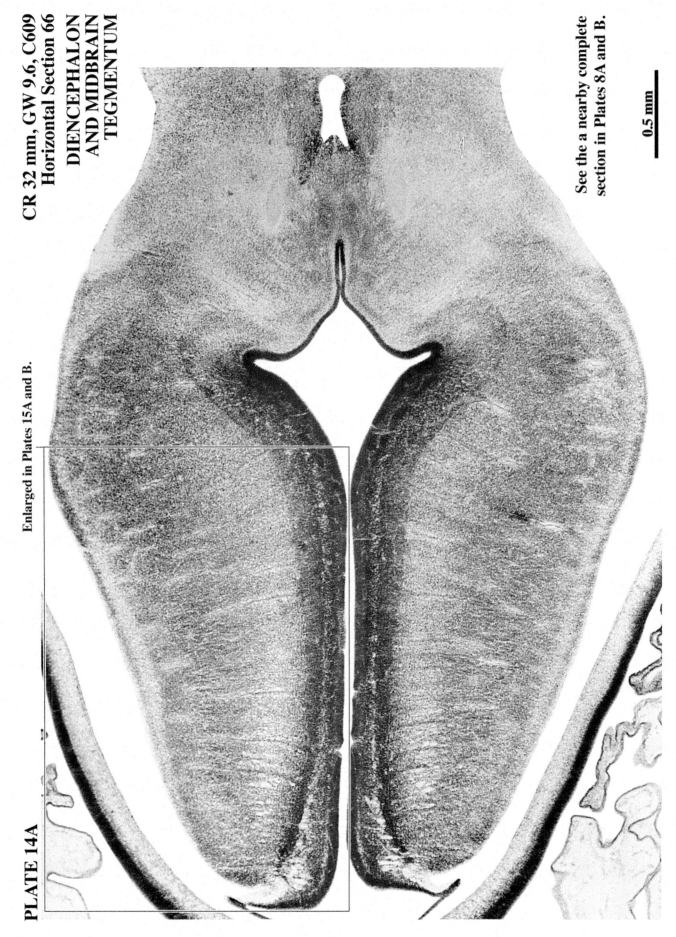

CR 32 mm, GW 9.6, C609
Horizontal Section 66
DIENCEPHALON
AND MIDBRAIN
TEGMENTUM

See the a nearby complete
section in Plates 8A and B.

0.5 mm

Enlarged in Plates 15A and B.

PLATE 14A

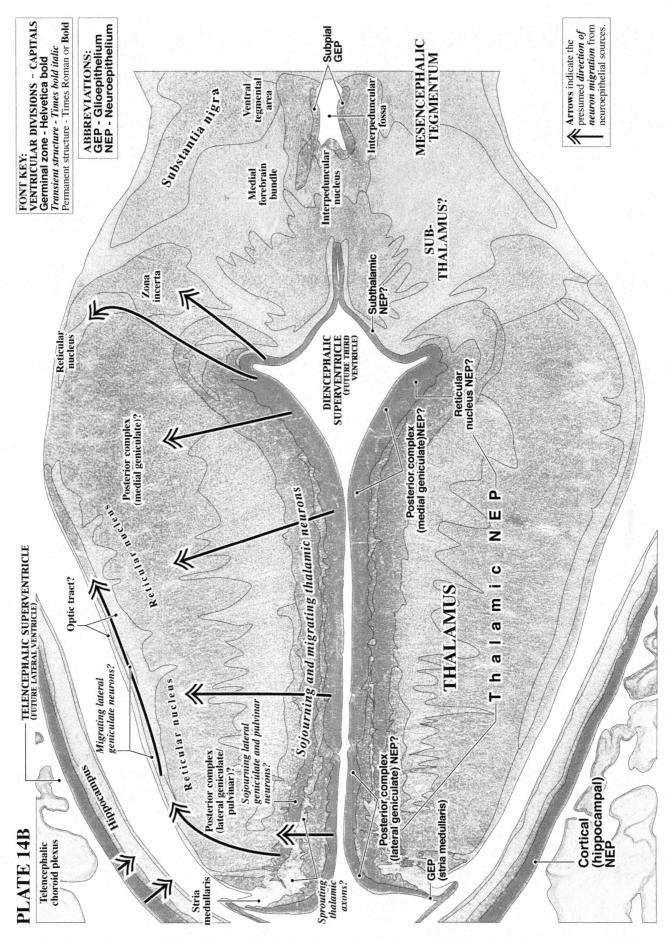

PLATE 14B

FONT KEY:
VENTRICULAR DIVISIONS – CAPITALS
Germinal zone - Helvetica bold
Transient structure - Times bold italic
Permanent structure - Times Roman or **Bold**

ABBREVIATIONS:
GEP - Glioepithelium
NEP - Neuroepithelium

Arrows indicate the
presumed *direction of
neuron migration* from
neuroepithelial sources.

Substantia nigra

Ventral
tegmental
area

Subpial
GEP

Medial
forebrain
bundle

Interpeduncular
fossa

Interpeduncular
nucleus

**MESENCEPHALIC
TEGMENTUM**

Subthalamic
NEP?

**SUB-
THALAMUS?**

Zona
incerta

Reticular
nucleus

*Posterior complex
(medial geniculate)?*

DIENCEPHALIC
SUPERVENTRICLE
(FUTURE THIRD
VENTRICLE)

Posterior complex
(medial geniculate) NEP?

R e t i c u l a r   n u c l e u s

Reticular
nucleus NEP?

TELENCEPHALIC SUPERVENTRICLE
(FUTURE LATERAL VENTRICLE)

Optic tract?

*Migrating lateral
geniculate neurons?*

R e t i c u l a r   n u c l e u s

*Posterior complex
(lateral geniculate/
pulvinar)?*

*Sojourning lateral
geniculate and pulvinar
neurons?*

Sojourning and migrating thalamic neurons

THALAMUS

T h a l a m i c   N E P

Posterior complex
(lateral geniculate) NEP?

Posterior complex
(lateral geniculate) NEP?

GEP
(stria medullaris)

Telencephalic
choroid plexus

Hippocampus

Stria
medullaris

*Sprouting
thalamic
axons?*

**Cortical
(hippocampal)
NEP**

48

PLATE 15A

CR 32 mm, GW 9.6, C609, Horizontal Section 66
DORSAL THALAMUS

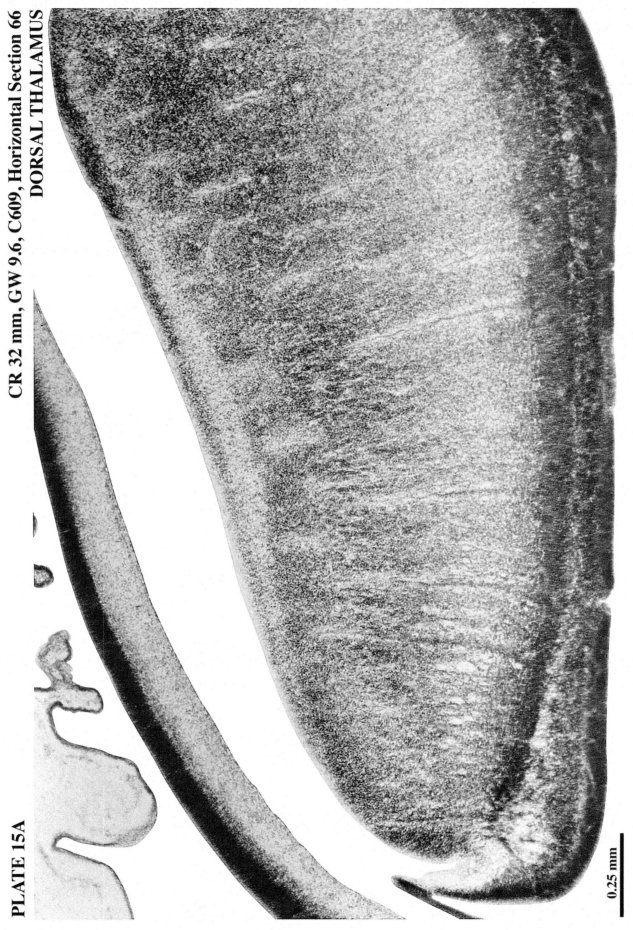

See the entire Section 65 in Plates 8A and B.

0.25 mm

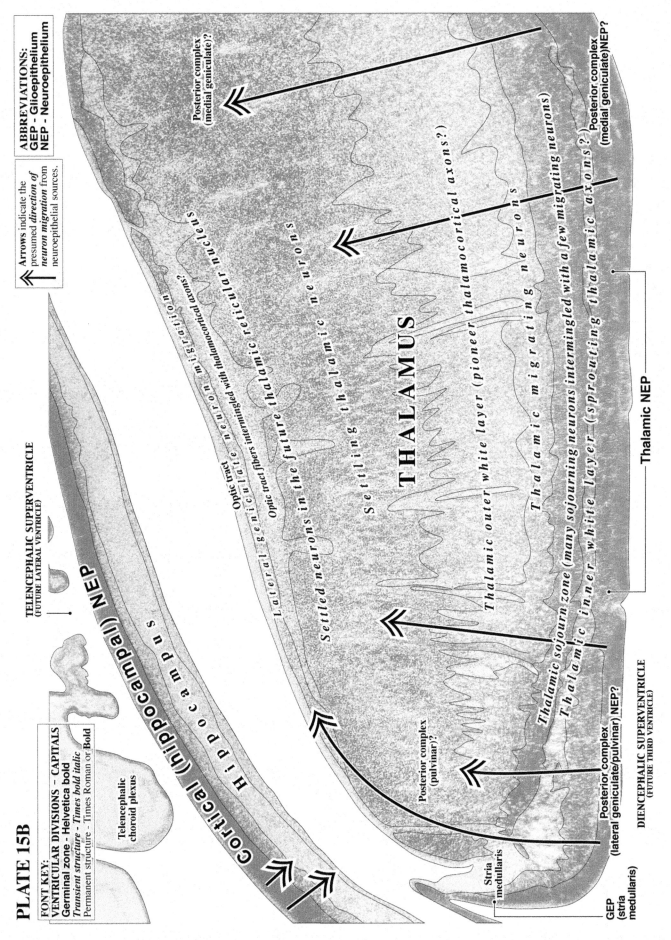

49

PLATE 15B

FONT KEY:
VENTRICULAR DIVISIONS - CAPITALS
Germinal zone - Helvetica bold
Transient structure - Times bold italic
Permanent structure - Times Roman or **Bold**

Telencephalic
choroid plexus

TELENCEPHALIC SUPERVENTRICLE
(FUTURE LATERAL VENTRICLE)

Arrows indicate the
presumed *direction of
neuron migration* from
neuroepithelial sources.

ABBREVIATIONS:
GEP - Glioepithelium
NEP - Neuroepithelium

Cortical (hippocampal) NEP

Hippocampus

Posterior complex
(medial geniculate)?

Optic tract neuron migration?

Lateral geniculate thalamocortical nuclei?

Optic tract fibers intermingled with thalamic reticular neurons?

Settled neurons in the future thalamic neurons

Settling thalamic neurons

THALAMUS

Thalamic outer white layer (pioneer thalamocortical axons?)

Thalamic migrating neurons

Posterior complex
(medial geniculate)NEP?

Thalamic sojourn zone (many sojourning neurons intermingled with a few migrating neurons)

Thalamic inner white layer (sprouting thalamic axons?)

Thalamic NEP

Posterior complex
(pulvinar)?

Posterior complex
(lateral geniculate/pulvinar) NEP?

Stria
medullaris

GEP
(stria
medullaris)

DIENCEPHALIC SUPERVENTRICLE
(FUTURE THIRD VENTRICLE)

50

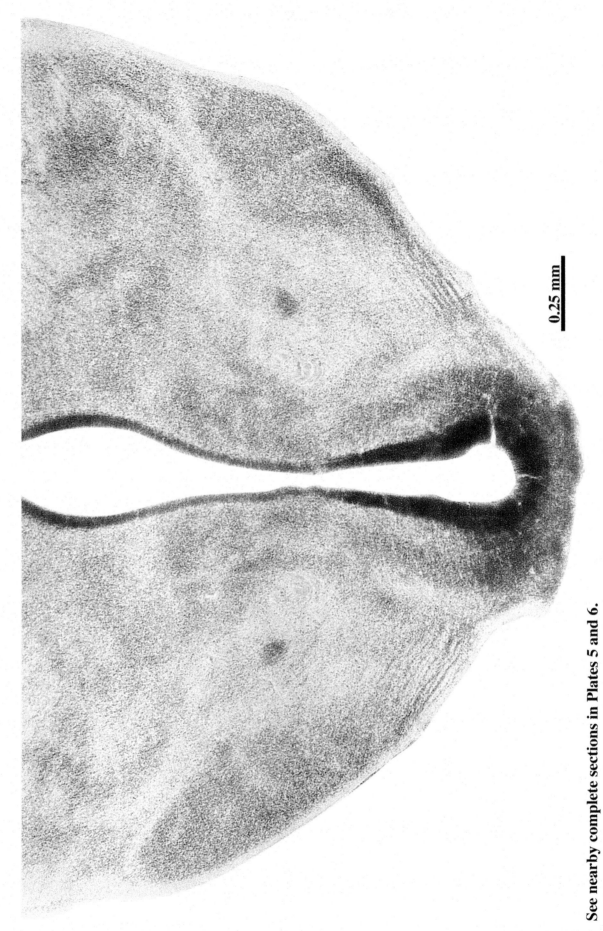

PLATE 16A

CR 32 mm, GW 9.6, C609, Horizontal Section 92
HYPOTHALAMUS AND SUBTHALAMUS

0.25 mm

See nearby complete sections in Plates 5 and 6.

PLATE 16B

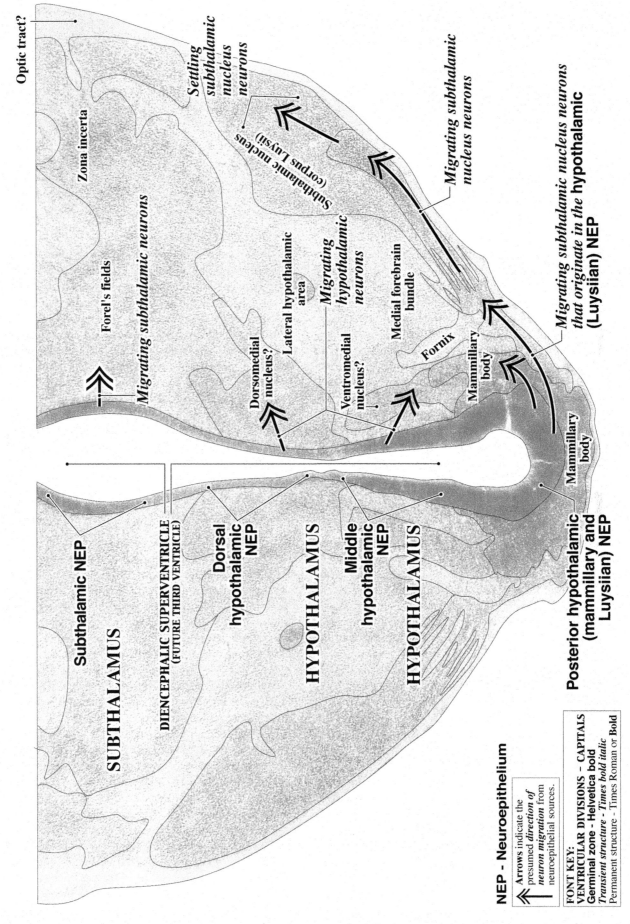

Optic tract?

Zona incerta

Settling subthalamic nucleus neurons

Migrating subthalamic nucleus neurons

Subthalamic nucleus (corpus Luysii)

Forel's fields

Migrating subthalamic neurons

Lateral hypothalamic area

Migrating hypothalamic neurons

Migrating subthalamic nucleus neurons that originate in the hypothalamic (Luysiian) NEP

Dorsomedial nucleus?

Ventromedial nucleus?

Medial forebrain bundle

Fornix

Mammillary body

Mammillary body

Subthalamic NEP

DIENCEPHALIC SUPERVENTRICLE (FUTURE THIRD VENTRICLE)

Dorsal hypothalamic NEP

SUBTHALAMUS

HYPOTHALAMUS

Middle hypothalamic NEP

HYPOTHALAMUS

Posterior hypothalamic (mammillary and Luysiian) NEP

NEP - Neuroepithelium

↞ **Arrows** indicate the presumed *direction of neuron migration* from neuroepithelial sources.

FONT KEY:
VENTRICULAR DIVISIONS - CAPITALS
Germinal zone - Helvetica bold
Transient structure - Times bold italic
Permanent structure - Times Roman or Bold

PART III: C9226
CR 31 mm (GW 9.5)
Frontal/Horizontal

Specimen C9226 from the Carnegie collection is a normal male fetus with a crown-rump length (CR) of 31 mm that was collected in 1954, and is estimated to be in gestational week (GW) 9.5. The entire fetus was embedded in paraffin mixed with 8% celloidin, cut transversely in 10-μm-thick sections, and stained with azan. The histology of this specimen is remarkable, and the sections are nearly perfectly bilateral. Since there is no photograph of this brain before it was embedded and cut, a specimen from Hochstetter (1919) that is only partially comparable to C9226 has been modified to show the approximate section plane and external features of the brain at GW8 (**Figure 15**). Like most of the specimens in this volume, the sections are not cut exactly in one plane; C9226's cortex is cut midway between coronal and horizontal planes. Since the cerebral cortex is in every section and the brainstem is cut in a more horizontal orientation, the brain more closely resembles a frontally sectioned brain. Unfortunately, the Hochstetter specimen is less mature (CR 27 mm) and we could not find a drawing of a brain specimen that would fit C9226. The C9226 sections through the cortex and brainstem are not in the same plane when transferred to Hochstetter's CR27 mm specimen. Instead, brainstem planes of sections appear to fan upward and downward from sections in the cortex. We interpret this to indicate that the brain flexures are more loosely folded in the Hochstetter specimen than in C9226. If one "squeezes" the brainstem to make the folds tighter, the cortex and brainstem planes would line up. Photographs of 23 sections are illustrated at low-magnification in **Plates 17-36**. High-magnification views of different areas of the cerebral cortex are shown in **Plates 37-38**.

C9226 is similar to the other GW9.5/9.6 specimens in this volume but shows the brain in a different perspective because of its unique cutting angle. The parenchyma is thick and bordered by a thin neuroepithelium (NEP) in the medulla, pons, and midbrain tegmentum, indicating that most neurons have been generated in these structures. Furthermore, the lack of dense accumulations of cells just outside the NEP in the midbrain tegmentum, pons, and medulla indicate that very few neurons are being generated, few are migrating, and most are settled and differentiating. The main exception is the **precerebellar neuroepithelium** in the medulla is thicker and generating pontine gray (and possibly other neurons); many precerebellar neurons are migrating in the **anterior and posterior extramural migratory streams**. The cerebellar NEP is thicker than that in the pons and medulla. The cerebellar parenchyma contains a very dense Purkinje cell sojourn zone outside the NEP and presumptive earlier-generated deep neurons lie in a superficial position. Like C609, the **external germinal layer (egl)** is barely visible emanating from the germinal trigone in the dorsal rhombic lip. The mesencephalic tectal NEP is thicker than the tegmental NEP and its very thin parenchyma contains dense sojourning and migrating tectal neurons adjacent to the NEP. The tectum is one of the most immature brain structures.

The diencephalic NEP is thicker indicating that many neurons are still being generated even though there is also a thick parenchyma, especially in the thalamus. That is because the thalamus is very large in the mature human brain. There are dense accumulations of young neurons in sojourn zones outside the hypothalamic and thalamic NEPs, indicating that cell migration is more active than final settling and differentiation.

Within the telencephalon, the cerebral cortex has a thick NEP and a very thin parenchyma, indicating that it is the most immature brain structure. The **stratified transitional field (STF)** contains **STF1** and **STF5** only in lateral areas. The pronounced anterolateral (thicker) to dorsomedial (thinner) maturation gradient is evident in both the **cortical plate** and the **STF** layers. In contrast, both the basal telencephalic NEP/SVZ and parenchyma are thick. That is because the basal telencephalon contains many early-generated neuronal populations (for example, globus pallidus and substantia innominata) and massive late-generated populations (striatal neurons in the caudate and putamen). Most of the neurons settling in the basal telencephalon at GW8 are those of the early-generated populations.

GW9.5 "FRONTAL/HORIZONTAL" SECTION PLANES

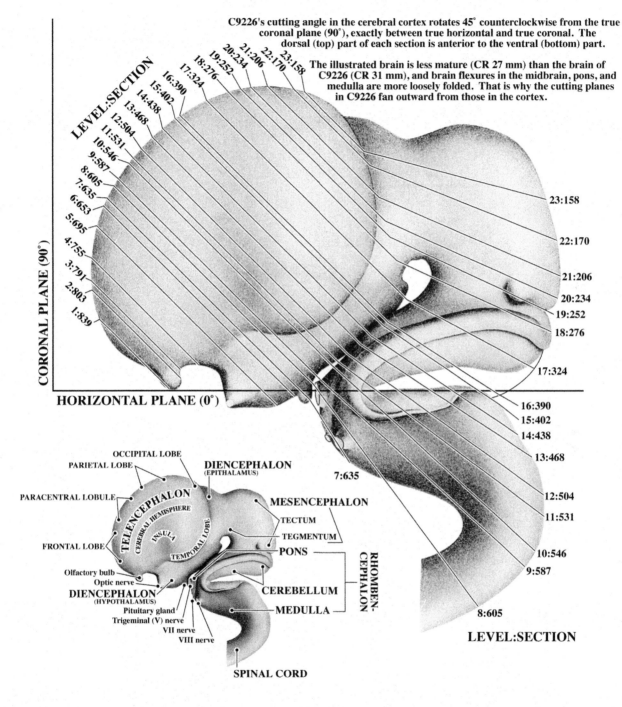

Figure 15. The lateral view of the brain and upper cervical spinal cord from a specimen with a crown-rump length of 27 mm (modified from Figure 37, Table VII, Hochstetter, 1919) serves to show the approximate locations and cutting angles of the illustrated sections of C9226 in the following pages. The small inset identifies the major structural features. The line in the cerebellum and dorsal edges of the pons and medulla is the cut edge of the medullary velum.

54

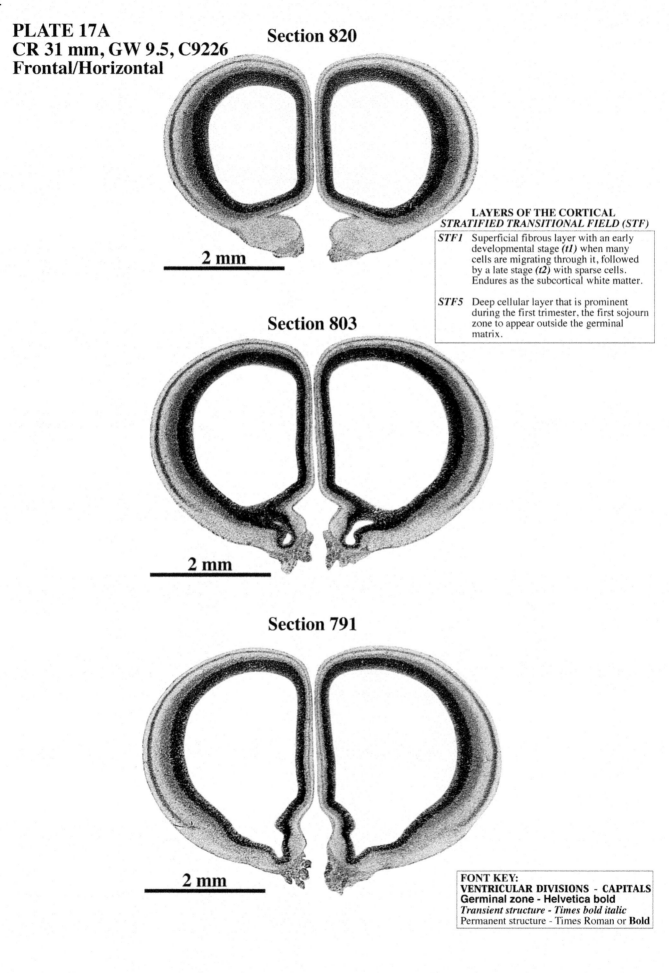

PLATE 17A
CR 31 mm, GW 9.5, C9226
Frontal/Horizontal

Section 820

2 mm

LAYERS OF THE CORTICAL
STRATIFIED TRANSITIONAL FIELD (STF)

STF1   Superficial fibrous layer with an early
developmental stage *(t1)* when many
cells are migrating through it, followed
by a late stage *(t2)* with sparse cells.
Endures as the subcortical white matter.

STF5   Deep cellular layer that is prominent
during the first trimester, the first sojourn
zone to appear outside the germinal
matrix.

Section 803

2 mm

Section 791

2 mm

FONT KEY:
VENTRICULAR DIVISIONS - CAPITALS
Germinal zone - Helvetica bold
Transient structure - Times bold italic
Permanent structure - Times Roman or **Bold**

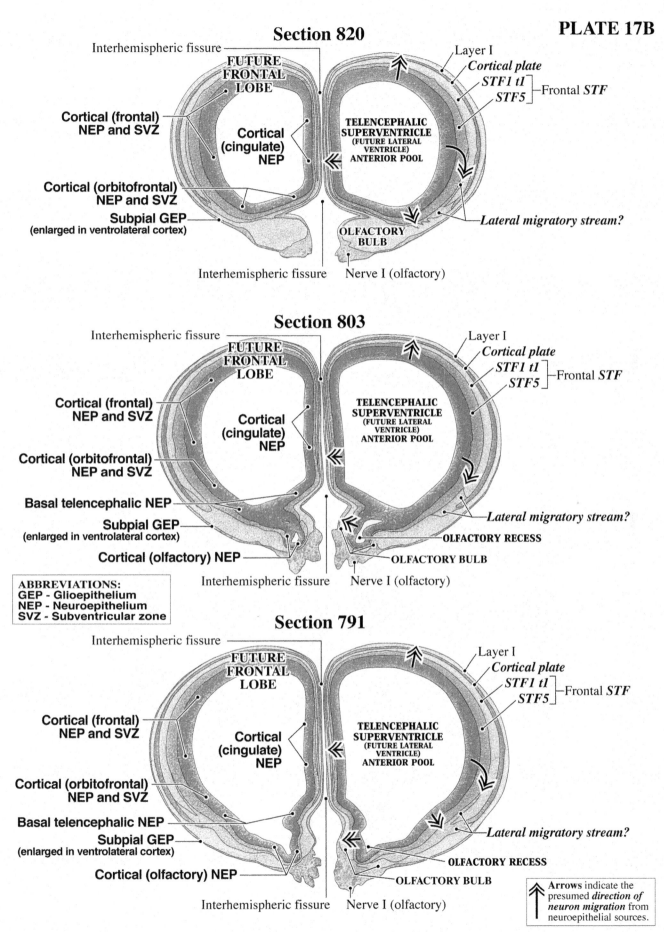

Section 820

Interhemispheric fissure

FUTURE
FRONTAL
LOBE

Layer I
Cortical plate
STF1 t1
STF5 —Frontal *STF*

Cortical (frontal)
NEP and SVZ

Cortical
(cingulate)
NEP

TELENCEPHALIC
SUPERVENTRICLE
(FUTURE LATERAL
VENTRICLE)
ANTERIOR POOL

Cortical (orbitofrontal)
NEP and SVZ

Subpial GEP
(enlarged in ventrolateral cortex)

Lateral migratory stream?

OLFACTORY
BULB

Interhemispheric fissure   Nerve I (olfactory)

Section 803

Interhemispheric fissure

FUTURE
FRONTAL
LOBE

Layer I
Cortical plate
STF1 t1
STF5 —Frontal *STF*

Cortical (frontal)
NEP and SVZ

Cortical
(cingulate)
NEP

TELENCEPHALIC
SUPERVENTRICLE
(FUTURE LATERAL
VENTRICLE)
ANTERIOR POOL

Cortical (orbitofrontal)
NEP and SVZ

Basal telencephalic NEP

Subpial GEP
(enlarged in ventrolateral cortex)

Lateral migratory stream?

OLFACTORY RECESS

Cortical (olfactory) NEP

OLFACTORY BULB

Interhemispheric fissure   Nerve I (olfactory)

ABBREVIATIONS:
GEP - Glioepithelium
NEP - Neuroepithelium
SVZ - Subventricular zone

Section 791

Interhemispheric fissure

FUTURE
FRONTAL
LOBE

Layer I
Cortical plate
STF1 t1
STF5 —Frontal *STF*

Cortical (frontal)
NEP and SVZ

Cortical
(cingulate)
NEP

TELENCEPHALIC
SUPERVENTRICLE
(FUTURE LATERAL
VENTRICLE)
ANTERIOR POOL

Cortical (orbitofrontal)
NEP and SVZ

Basal telencephalic NEP

Subpial GEP
(enlarged in ventrolateral cortex)

Lateral migratory stream?

Cortical (olfactory) NEP

OLFACTORY RECESS

OLFACTORY BULB

Interhemispheric fissure   Nerve I (olfactory)

Arrows indicate the
presumed *direction of
neuron migration* from
neuroepithelial sources.

PLATE 18A
CR 31 mm, GW 9.5, C9226
Frontal/Horizontal

Section 755

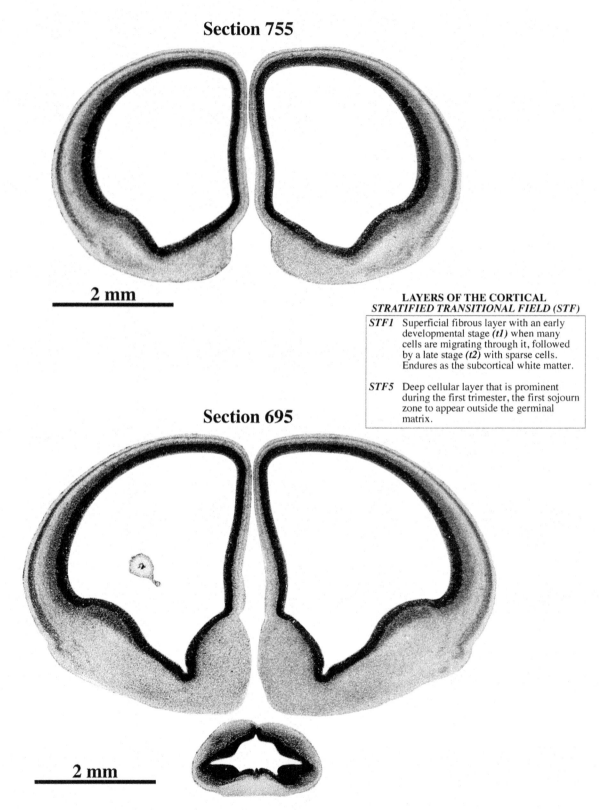

2 mm

Section 695

LAYERS OF THE CORTICAL
STRATIFIED TRANSITIONAL FIELD (STF)

STF1	Superficial fibrous layer with an early developmental stage *(t1)* when many cells are migrating through it, followed by a late stage *(t2)* with sparse cells. Endures as the subcortical white matter.
STF5	Deep cellular layer that is prominent during the first trimester, the first sojourn zone to appear outside the germinal matrix.

2 mm

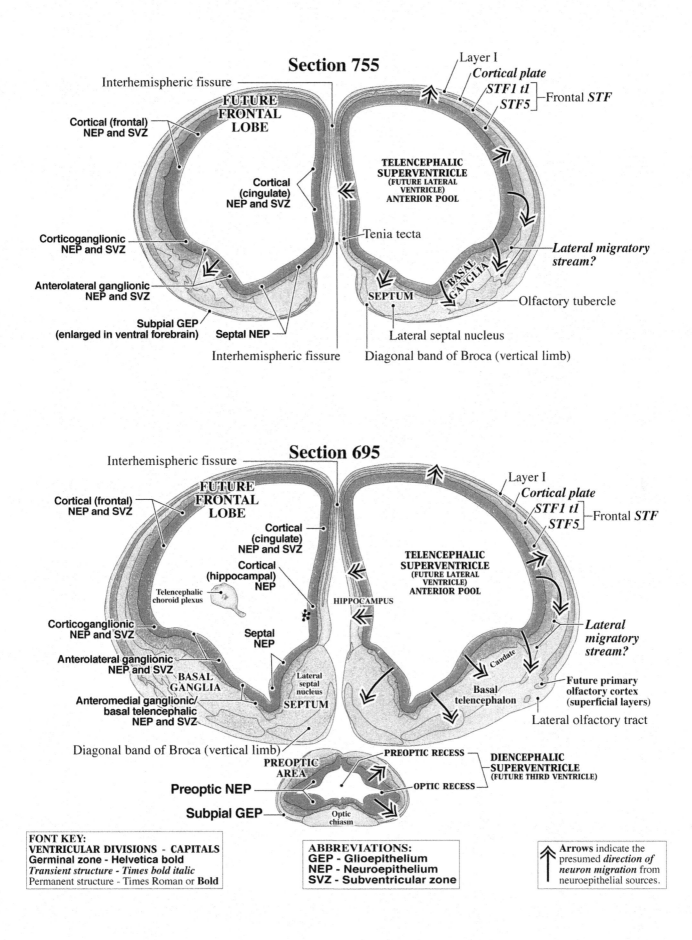

Section 755

Interhemispheric fissure

Layer I
Cortical plate
STF1 t1
STF5 ⎤ Frontal *STF*

FUTURE
FRONTAL
LOBE

Cortical (frontal)
NEP and SVZ

TELENCEPHALIC
SUPERVENTRICLE
(FUTURE LATERAL
VENTRICLE)
ANTERIOR POOL

Cortical
(cingulate)
NEP and SVZ

Tenia tecta

*Lateral migratory
stream?*

BASAL
GANGLIA

Corticoganglionic
NEP and SVZ

SEPTUM

Olfactory tubercle

Anterolateral ganglionic
NEP and SVZ

Subpial GEP
(enlarged in ventral forebrain) Septal NEP

Lateral septal nucleus

Interhemispheric fissure

Diagonal band of Broca (vertical limb)

Section 695

Interhemispheric fissure

FUTURE
FRONTAL
LOBE

Layer I
Cortical plate
STF1 t1
STF5 ⎤ Frontal *STF*

Cortical (frontal)
NEP and SVZ

Cortical
(cingulate)
NEP and SVZ

Cortical
(hippocampal)
NEP

TELENCEPHALIC
SUPERVENTRICLE
(FUTURE LATERAL
VENTRICLE)
ANTERIOR POOL

Telencephalic
choroid plexus

HIPPOCAMPUS

Corticoganglionic
NEP and SVZ

Septal
NEP

*Lateral
migratory
stream?*

Anterolateral ganglionic
NEP and SVZ

BASAL
GANGLIA

Lateral
septal
nucleus

Caudate

Future primary
olfactory cortex
(superficial layers)

Anteromedial ganglionic/
basal telencephalic
NEP and SVZ

SEPTUM

Basal
telencephalon

Lateral olfactory tract

Diagonal band of Broca (vertical limb)

PREOPTIC RECESS

PREOPTIC
AREA

DIENCEPHALIC
SUPERVENTRICLE
(FUTURE THIRD VENTRICLE)

Preoptic NEP

OPTIC RECESS

Subpial GEP

Optic
chiasm

FONT KEY:
VENTRICULAR DIVISIONS - CAPITALS
Germinal zone - Helvetica bold
Transient structure - Times bold italic
Permanent structure - Times Roman or **Bold**

ABBREVIATIONS:
GEP - Glioepithelium
NEP - Neuroepithelium
SVZ - Subventricular zone

Arrows indicate the
presumed *direction of
neuron migration* from
neuroepithelial sources.

58

PLATE 19A
CR 31 mm, GW 9.5, C9226
Frontal/Horizontal, Section 653

LAYERS OF THE CORTICAL
STRATIFIED TRANSITIONAL FIELD (STF)

STF1 Superficial fibrous layer with an early developmental stage *(t1)* when many cells are migrating through it, followed by a late stage *(t2)* with sparse cells. Endures as the subcortical white matter.

STF5 Deep cellular layer that is prominent during the first trimester, the first sojourn zone to appear outside the germinal matrix.

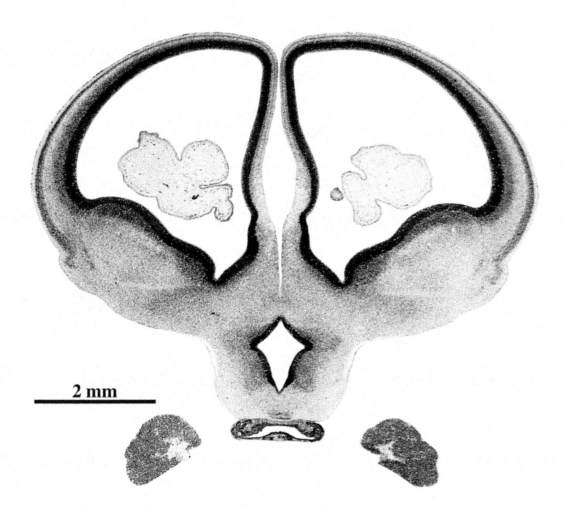

2 mm

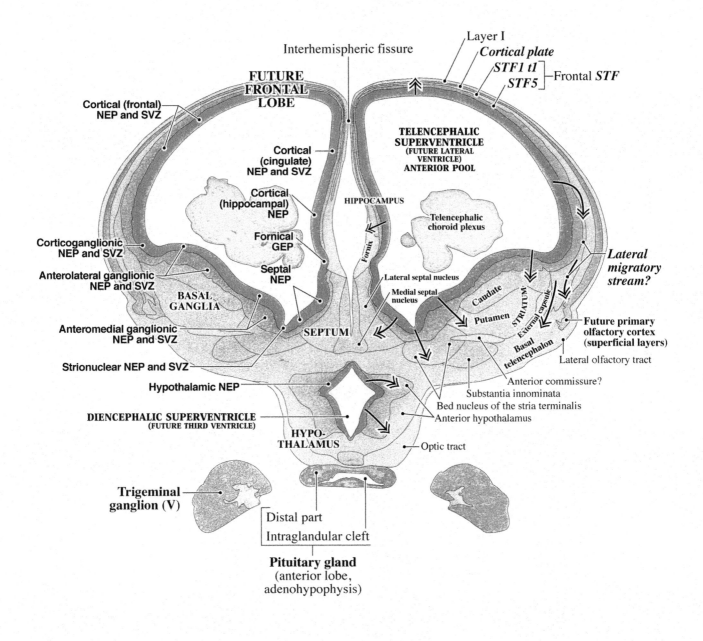

Layer I
Cortical plate
STF1 t1
STF5 ⎤Frontal *STF*

Interhemispheric fissure

FUTURE FRONTAL LOBE

Cortical (frontal) NEP and SVZ

Cortical (cingulate) NEP and SVZ

Cortical (hippocampal) NEP

Fornical GEP

Septal NEP

TELENCEPHALIC SUPERVENTRICLE (FUTURE LATERAL VENTRICLE) **ANTERIOR POOL**

HIPPOCAMPUS

Telencephalic choroid plexus

Fornix

Corticoganglionic NEP and SVZ

Anterolateral ganglionic NEP and SVZ

BASAL GANGLIA

Lateral septal nucleus

Medial septal nucleus

Caudate

Putamen

STRIATUM

External capsule

Lateral migratory stream?

Anteromedial ganglionic NEP and SVZ

SEPTUM

Basal telencephalon

Future primary olfactory cortex (superficial layers)

Lateral olfactory tract

Strionuclear NEP and SVZ

Hypothalamic NEP

Anterior commissure?

Substantia innominata

Bed nucleus of the stria terminalis

Anterior hypothalamus

DIENCEPHALIC SUPERVENTRICLE (FUTURE THIRD VENTRICLE)

HYPO-THALAMUS

Optic tract

Trigeminal ganglion (V)

Distal part

Intraglandular cleft

Pituitary gland (anterior lobe, adenohypophysis)

FONT KEY:
VENTRICULAR DIVISIONS - CAPITALS
Germinal zone - Helvetica bold
Transient structure - Times bold italic
Permanent structure - Times Roman or **Bold**

ABBREVIATIONS:
GEP - Glioepithelium
NEP - Neuroepithelium
SVZ - Subventricular zone

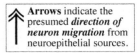

Arrows indicate the presumed *direction of neuron migration* from neuroepithelial sources.

PLATE 20A
CR 31 mm, GW 9.5, C9226
Frontal/Horizontal, Section 635

LAYERS OF THE CORTICAL
STRATIFIED TRANSITIONAL FIELD (STF)

STF1	Superficial fibrous layer with an early developmental stage *(t1)* when many cells are migrating through it, followed by a late stage *(t2)* with sparse cells. Endures as the subcortical white matter.
STF5	Deep cellular layer that is prominent during the first trimester, the first sojourn zone to appear outside the germinal matrix.

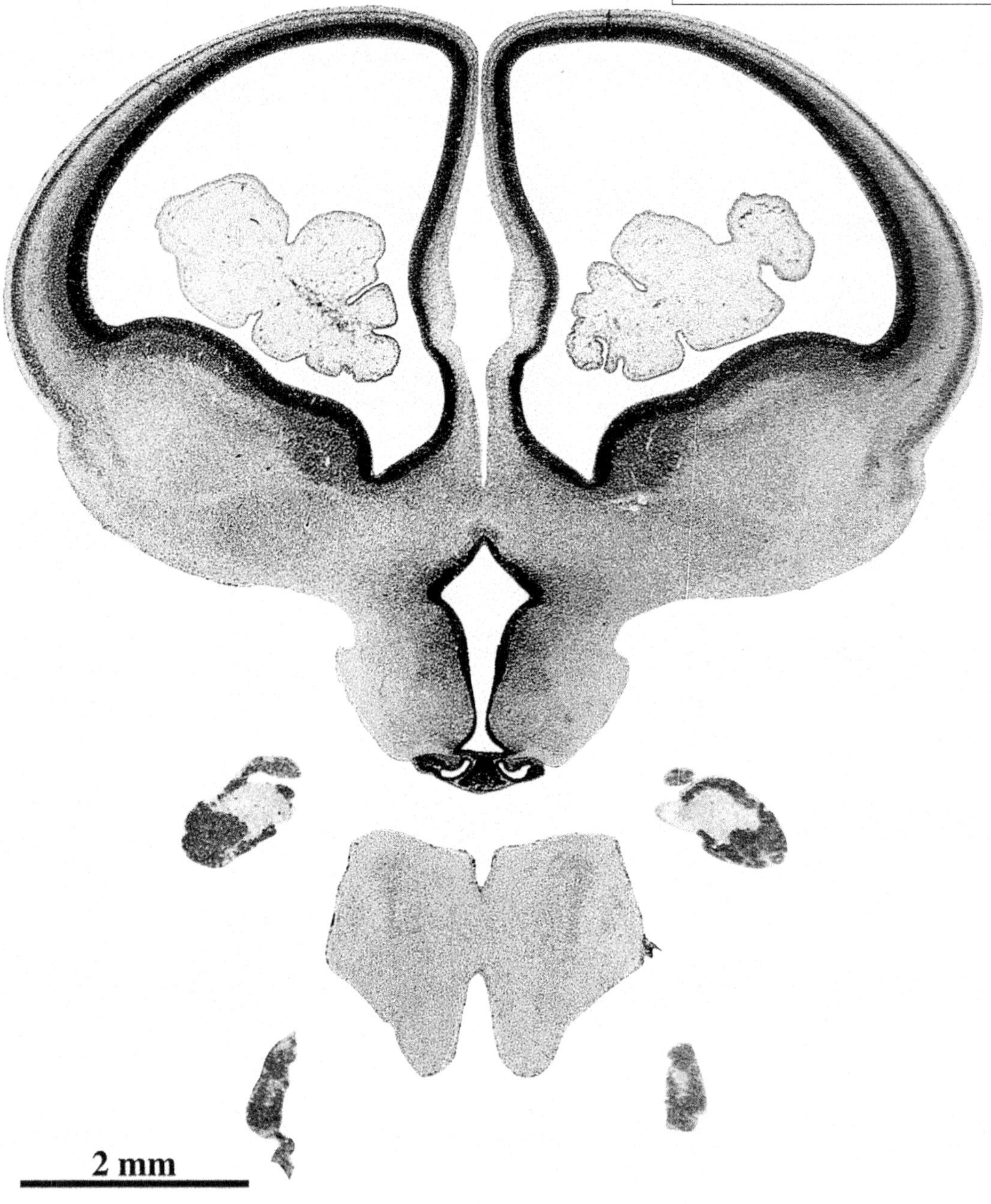

2 mm

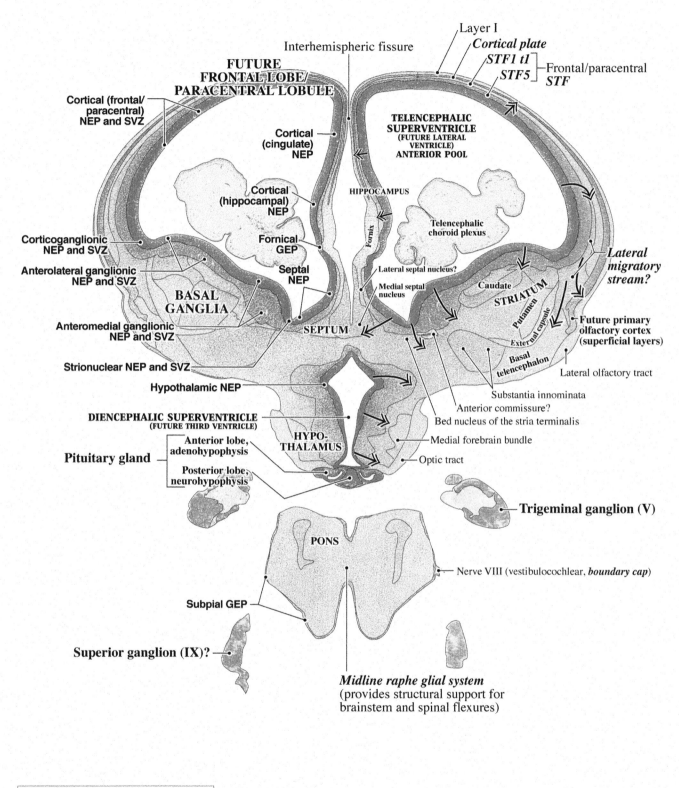

Interhemispheric fissure

Layer I
Cortical plate
STF1 t1
STF5 } Frontal/paracentral *STF*

FUTURE FRONTAL LOBE/ PARACENTRAL LOBULE

Cortical (frontal/ paracentral) NEP and SVZ

Cortical (cingulate) NEP

TELENCEPHALIC SUPERVENTRICLE (FUTURE LATERAL VENTRICLE) ANTERIOR POOL

Cortical (hippocampal) NEP

HIPPOCAMPUS

Telencephalic choroid plexus

Fornix

Fornical GEP

Corticoganglionic NEP and SVZ

Anterolateral ganglionic NEP and SVZ

Septal NEP

Lateral septal nucleus?

Medial septal nucleus

BASAL GANGLIA

Caudate

STRIATUM

Lateral migratory stream?

Putamen

Anteromedial ganglionic NEP and SVZ

SEPTUM

External capsule

Future primary olfactory cortex (superficial layers)

Strionuclear NEP and SVZ

Basal telencephalon

Lateral olfactory tract

Hypothalamic NEP

Substantia innominata

DIENCEPHALIC SUPERVENTRICLE (FUTURE THIRD VENTRICLE)

Anterior commissure?

Bed nucleus of the stria terminalis

Pituitary gland

Anterior lobe, adenohypophysis

HYPO-THALAMUS

Medial forebrain bundle

Posterior lobe, neurohypophysis

Optic tract

Trigeminal ganglion (V)

PONS

Nerve VIII (vestibulocochlear, *boundary cap*)

Subpial GEP

Superior ganglion (IX)?

Midline raphe glial system
(provides structural support for brainstem and spinal flexures)

FONT KEY:
VENTRICULAR DIVISIONS - CAPITALS
Germinal zone - Helvetica bold
Transient structure - Times bold italic
Permanent structure - Times Roman or **Bold**

ABBREVIATIONS:
GEP - Glioepithelium
NEP - Neuroepithelium
SVZ - Subventricular zone

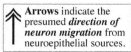

Arrows indicate the presumed *direction of neuron migration* from neuroepithelial sources.

**PLATE 21A
CR 31 mm, GW 9.5, C9226
Frontal/Horizontal, Section 605**

**LAYERS OF THE CORTICAL
*STRATIFIED TRANSITIONAL FIELD (STF)***

STF1 Superficial fibrous layer with an early
developmental stage *(t1)* when many
cells are migrating through it, followed
by a late stage *(t2)* with sparse cells.
Endures as the subcortical white matter.

STF5 Deep cellular layer that is prominent
during the first trimester, the first sojourn
zone to appear outside the germinal
matrix.

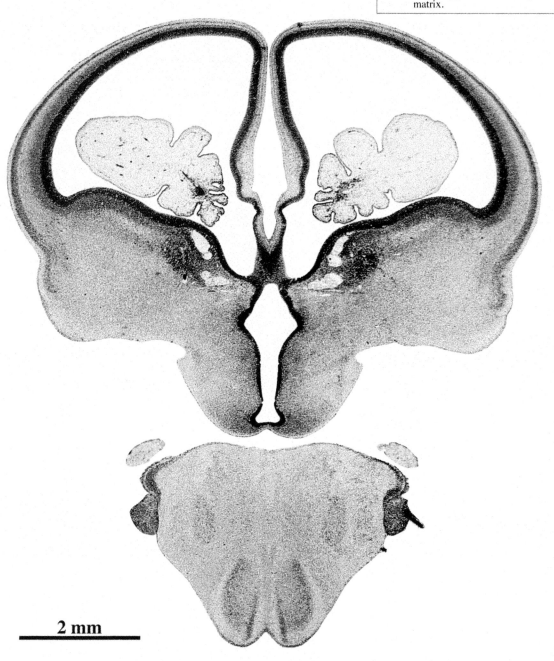

2 mm

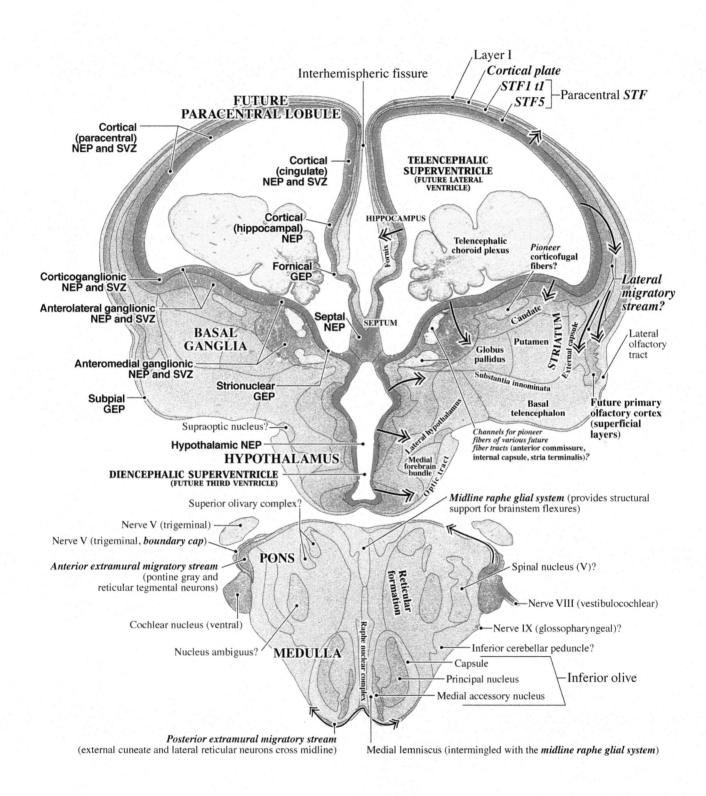

Layer I
Cortical plate
STF1 t1
STF5]—Paracentral *STF*

Interhemispheric fissure

**FUTURE
PARACENTRAL LOBULE**

**TELENCEPHALIC
SUPERVENTRICLE**
(FUTURE LATERAL
VENTRICLE)

**Cortical
(paracentral)
NEP and SVZ**

**Cortical
(cingulate)
NEP and SVZ**

HIPPOCAMPUS

**Cortical
(hippocampal)
NEP**

Telencephalic
choroid plexus

*Pioneer
corticofugal
fibers?*

*Lateral
migratory
stream?*

**Fornical
GEP**

Caudate

**Corticoganglionic
NEP and SVZ**

**Anterolateral ganglionic
NEP and SVZ**

**Septal
NEP**

SEPTUM

Putamen

STRIATUM

External capsule

Lateral
olfactory
tract

**BASAL
GANGLIA**

Globus
pallidus

**Anteromedial ganglionic
NEP and SVZ**

Substantia innominata

**Future primary
olfactory cortex
(superficial
layers)**

**Strionuclear
GEP**

**Subpial
GEP**

Basal
telencephalon

Supraoptic nucleus?

*Channels for pioneer
fibers of various future
fiber tracts (anterior commissure,
internal capsule, stria terminalis)?*

Hypothalamic NEP

Lateral hypothalamus

HYPOTHALAMUS

*Medial
forebrain
bundle*

Optic tract

DIENCEPHALIC SUPERVENTRICLE
(FUTURE THIRD VENTRICLE)

Midline raphe glial system (provides structural
support for brainstem flexures)

Superior olivary complex?

Nerve V (trigeminal)

Nerve V (trigeminal, *boundary cap*)

*Reticular
formation*

PONS

Spinal nucleus (V)?

Anterior extramural migratory stream
(pontine gray and
reticular tegmental neurons)

Nerve VIII (vestibulocochlear)

Nerve IX (glossopharyngeal)?

Cochlear nucleus (ventral)

Raphe nuclear complex

Inferior cerebellar peduncle?

Nucleus ambiguus?

MEDULLA

Capsule

Principal nucleus

Medial accessory nucleus

—Inferior olive

Posterior extramural migratory stream
(external cuneate and lateral reticular neurons cross midline)

Medial lemniscus (intermingled with the *midline raphe glial system*)

FONT KEY:
VENTRICULAR DIVISIONS - CAPITALS
Germinal zone - Helvetica bold
Transient structure - Times bold italic
Permanent structure - Times Roman or **Bold**

ABBREVIATIONS:
GEP - Glioepithelium
NEP - Neuroepithelium
SVZ - Subventricular zone

Arrows indicate the
presumed *direction of
neuron migration* from
neuroepithelial sources.

PLATE 22A
CR 31 mm, GW 9.5, C9226
Frontal/Horizontal, Section 587

**See high-magnification views of
the cerebral cortex freom
nearby sections in Plates
37 to 38A and B.**

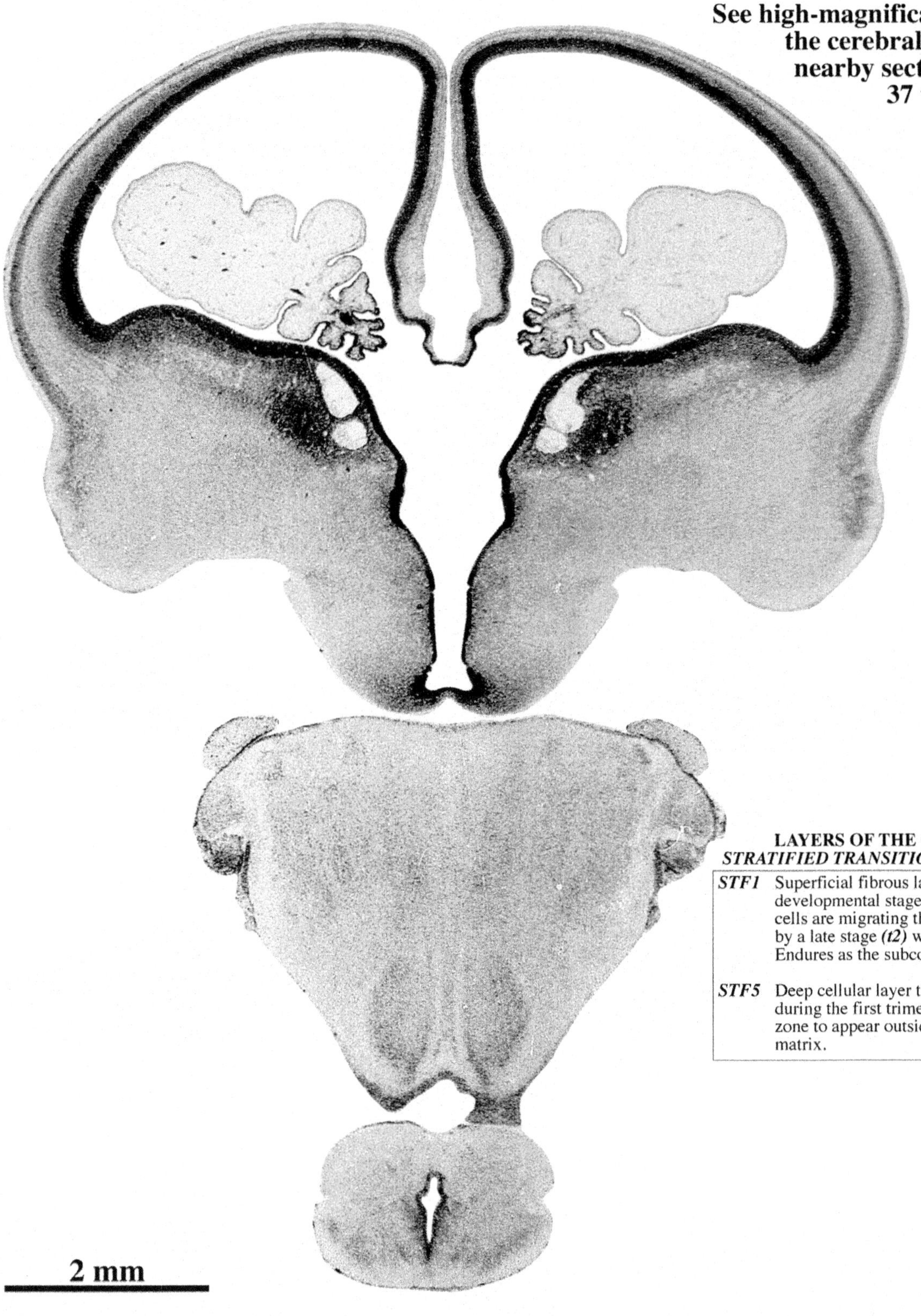

2 mm

LAYERS OF THE CORTICAL
STRATIFIED TRANSITIONAL FIELD (STF)

STF1	Superficial fibrous layer with an early developmental stage *(t1)* when many cells are migrating through it, followed by a late stage *(t2)* with sparse cells. Endures as the subcortical white matter.
STF5	Deep cellular layer that is prominent during the first trimester, the first sojourn zone to appear outside the germinal matrix.

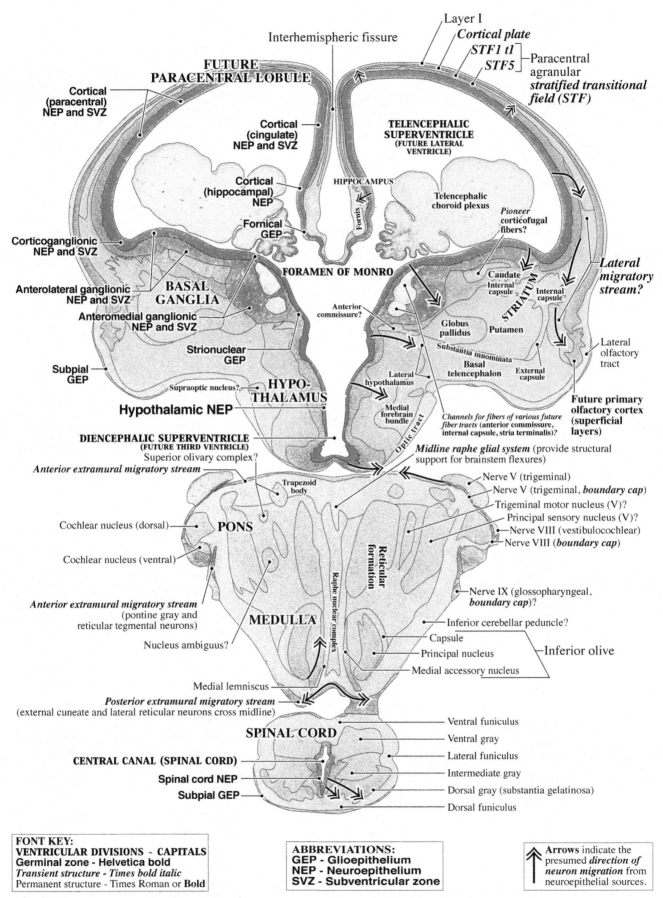

Interhemispheric fissure

Layer I
Cortical plate
STF1 t1
STF5
Paracentral
agranular
*stratified transitional
field (STF)*

**FUTURE
PARACENTRAL LOBULE**

Cortical
(paracentral)
NEP and SVZ

Cortical
(cingulate)
NEP and SVZ

**TELENCEPHALIC
SUPERVENTRICLE
(FUTURE LATERAL
VENTRICLE)**

Cortical
(hippocampal)
NEP

HIPPOCAMPUS

Telencephalic
choroid plexus

*Pioneer
corticofugal
fibers?*

Fornical
GEP

Fornix

Corticoganglionic
NEP and SVZ

FORAMEN OF MONRO

Caudate
Internal
capsule

Internal
capsule?

*Lateral
migratory
stream?*

Anterolateral ganglionic
NEP and SVZ

**BASAL
GANGLIA**

Anterior
commissure?

Globus
pallidus

Putamen

STRIATUM

Anteromedial ganglionic
NEP and SVZ

Strionuclear
GEP

Substantia innominata

**Basal
telencephalon**

External
capsule

Lateral
olfactory
tract

**Subpial
GEP**

Lateral
hypothalamus

Supraoptic nucleus?

**HYPO-
THALAMUS**

Medial
forebrain
bundle

*Channels for fibers of various future
fiber tracts (anterior commissure,
internal capsule, stria terminalis)?*

**Future primary
olfactory cortex
(superficial
layers)**

Hypothalamic NEP

Optic tract

DIENCEPHALIC SUPERVENTRICLE
(FUTURE THIRD VENTRICLE)

Superior olivary complex?

Midline raphe glial system (provide structural
support for brainstem flexures)

Anterior extramural migratory stream

Trapezoid
body

Nerve V (trigeminal)
Nerve V (trigeminal, *boundary cap*)
Trigeminal motor nucleus (V)?
Principal sensory nucleus (V)?
Nerve VIII (vestibulocochlear)
Nerve VIII (*boundary cap*)

Cochlear nucleus (dorsal)

PONS

*Reticular
formation*

Cochlear nucleus (ventral)

Raphe nuclear complex

Nerve IX (glossopharyngeal,
boundary cap)?

Anterior extramural migratory stream
(pontine gray and
reticular tegmental neurons)

MEDULLA

Inferior cerebellar peduncle?

Capsule
Principal nucleus

Inferior olive

Nucleus ambiguus?

Medial accessory nucleus

Medial lemniscus

Posterior extramural migratory stream
(external cuneate and lateral reticular neurons cross midline)

SPINAL CORD

Ventral funiculus
Ventral gray
Lateral funiculus

CENTRAL CANAL (SPINAL CORD)

Intermediate gray

Spinal cord NEP

Dorsal gray (substantia gelatinosa)

Subpial GEP

Dorsal funiculus

FONT KEY:
VENTRICULAR DIVISIONS - CAPITALS
Germinal zone - Helvetica bold
Transient structure - Times bold italic
Permanent structure - Times Roman or **Bold**

ABBREVIATIONS:
GEP - Glioepithelium
NEP - Neuroepithelium
SVZ - Subventricular zone

Arrows indicate the
presumed *direction of
neuron migration* from
neuroepithelial sources.

PLATE 23A
CR 31 mm, GW 9.5, C9226
Frontal/Horizontal, Section 546

See high-magnification views of
the cerebral cortex from
nearby sections in Plates
37 to 38A and B.

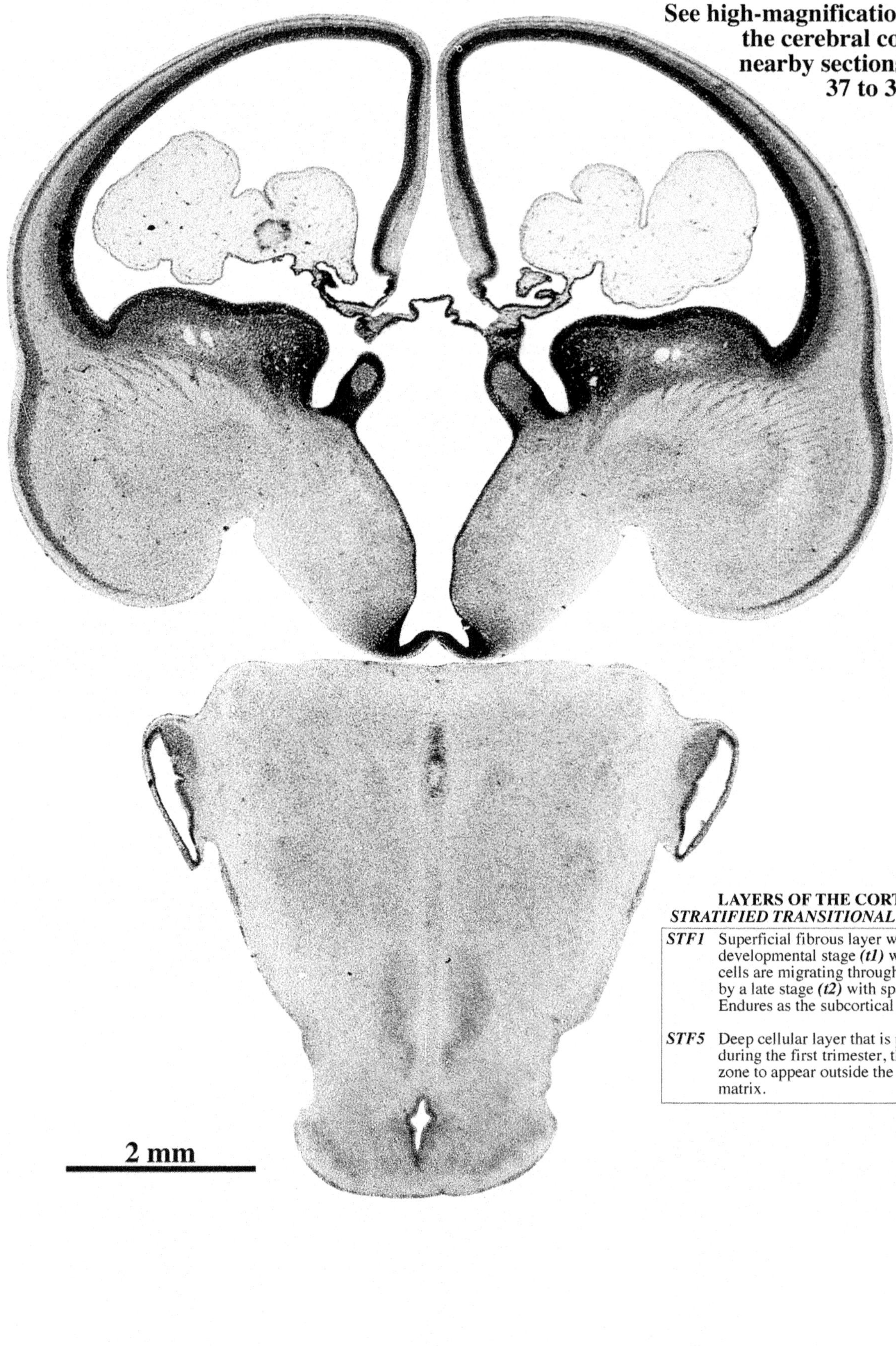

2 mm

LAYERS OF THE CORTICAL
STRATIFIED TRANSITIONAL FIELD (STF)

STF1 Superficial fibrous layer with an early
developmental stage *(t1)* when many
cells are migrating through it, followed
by a late stage *(t2)* with sparse cells.
Endures as the subcortical white matter.

STF5 Deep cellular layer that is prominent
during the first trimester, the first sojourn
zone to appear outside the germinal
matrix.

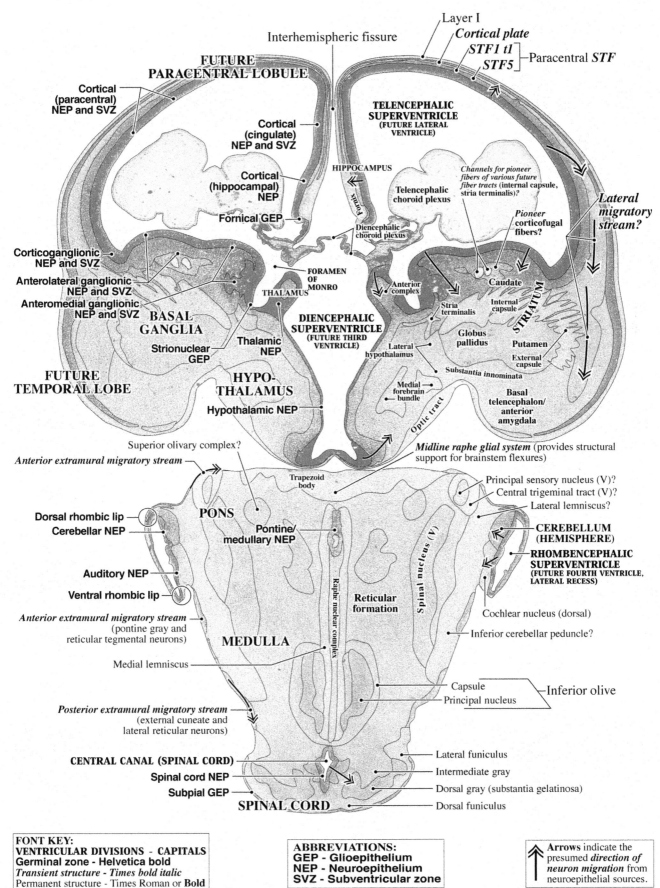

FUTURE
PARACENTRAL LOBULE

Interhemispheric fissure

Layer I
Cortical plate
STF1 t1
STF5 Paracentral *STF*

Cortical
(paracentral)
NEP and SVZ

Cortical
(cingulate)
NEP and SVZ

TELENCEPHALIC
SUPERVENTRICLE
(FUTURE LATERAL
VENTRICLE)

Cortical
(hippocampal)
NEP

HIPPOCAMPUS

*Channels for pioneer
fibers of various future
fiber tracts (internal capsule,
stria terminalis)?*

Telencephalic
choroid plexus

*Pioneer
corticofugal
fibers?*

Fornical GEP

*Lateral
migratory
stream?*

Diencephalic
choroid plexus

Fornix

Corticoganglionic
NEP and SVZ

FORAMEN
OF
MONRO

Anterior
complex

Caudate

STRIATUM

Anterolateral ganglionic
NEP and SVZ

THALAMUS

Stria
terminalis

Internal
capsule

Anteromedial ganglionic
NEP and SVZ

BASAL
GANGLIA

DIENCEPHALIC
SUPERVENTRICLE
(FUTURE THIRD
VENTRICLE)

Globus
pallidus

Putamen

Strionuclear
GEP

Thalamic
NEP

Lateral
hypothalamus

External
capsule

FUTURE
TEMPORAL LOBE

HYPO-
THALAMUS

Substantia innominata

Basal
telencephalon/
anterior
amygdala

Hypothalamic NEP

Medial
forebrain
bundle

Optic tract

Superior olivary complex?

Midline raphe glial system (provides structural
support for brainstem flexures)

Anterior extramural migratory stream

Trapezoid
body

Principal sensory nucleus (V)?
Central trigeminal tract (V)?
Lateral lemniscus?

PONS

Dorsal rhombic lip
Cerebellar NEP

Pontine/
medullary NEP

Spinal nucleus (V)

CEREBELLUM
(HEMISPHERE)

RHOMBENCEPHALIC
SUPERVENTRICLE
(FUTURE FOURTH VENTRICLE,
LATERAL RECESS)

Auditory NEP

Ventral rhombic lip

Raphe nuclear complex

**Reticular
formation**

Cochlear nucleus (dorsal)

Anterior extramural migratory stream
(pontine gray and
reticular tegmental neurons)

MEDULLA

Inferior cerebellar peduncle?

Medial lemniscus

Capsule

Inferior olive

Principal nucleus

Posterior extramural migratory stream
(external cuneate and
lateral reticular neurons)

Lateral funiculus

CENTRAL CANAL (SPINAL CORD)

Intermediate gray

Spinal cord NEP

Dorsal gray (substantia gelatinosa)

Subpial GEP

SPINAL CORD

Dorsal funiculus

FONT KEY:
VENTRICULAR DIVISIONS - CAPITALS
Germinal zone - Helvetica bold
Transient structure - Times bold italic
Permanent structure - Times Roman or **Bold**

ABBREVIATIONS:
GEP - Glioepithelium
NEP - Neuroepithelium
SVZ - Subventricular zone

Arrows indicate the
presumed *direction of
neuron migration* from
neuroepithelial sources.

PLATE 24A
CR 31 mm, GW 9.5, C9226
Frontal/Horizontal, Section 531

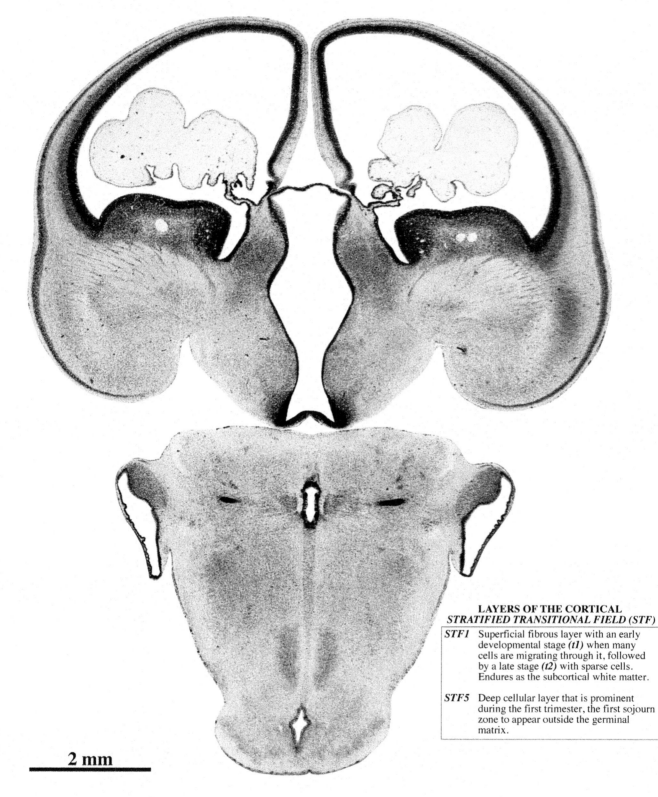

2 mm

LAYERS OF THE CORTICAL
STRATIFIED TRANSITIONAL FIELD (STF)

STF1 Superficial fibrous layer with an early
developmental stage *(t1)* when many
cells are migrating through it, followed
by a late stage *(t2)* with sparse cells.
Endures as the subcortical white matter.

STF5 Deep cellular layer that is prominent
during the first trimester, the first sojourn
zone to appear outside the germinal
matrix.

PLATE 24B

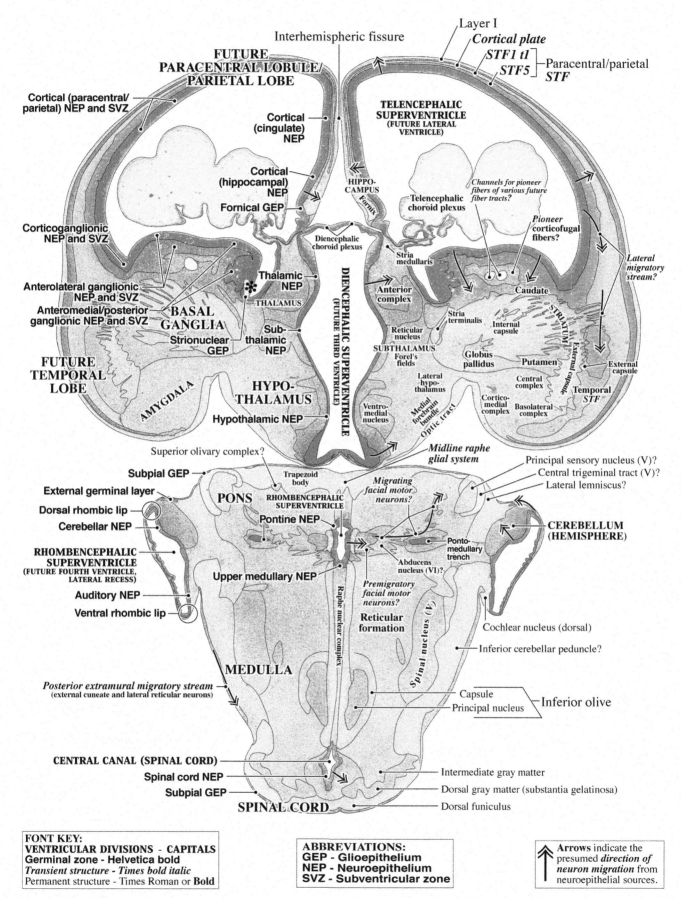

Layer I
Cortical plate
STF1 tl
STF5
Paracentral/parietal *STF*

Interhemispheric fissure

FUTURE PARACENTRAL LOBULE/ PARIETAL LOBE

Cortical (paracentral/ parietal) NEP and SVZ

Cortical (cingulate) NEP

TELENCEPHALIC SUPERVENTRICLE (FUTURE LATERAL VENTRICLE)

Cortical (hippocampal) NEP

HIPPO-CAMPUS

Fornix

Channels for pioneer fibers of various future fiber tracts?

Telencephalic choroid plexus

Pioneer corticofugal fibers?

Fornical GEP

Diencephalic choroid plexus

Stria medullaris

Corticoganglionic NEP and SVZ

Thalamic NEP

THALAMUS

DIENCEPHALIC SUPERVENTRICLE (FUTURE THIRD VENTRICLE)

Anterior complex

Caudate

Lateral migratory stream?

Anterolateral ganglionic NEP and SVZ

Anteromedial/posterior ganglionic NEP and SVZ

BASAL GANGLIA

Strionuclear GEP

Sub-thalamic NEP

Stria terminalis

Reticular nucleus

STRIATUM

Internal capsule

External capsule

FUTURE TEMPORAL LOBE

SUBTHALAMUS
Forel's fields

Globus pallidus

Putamen

External capsule

HYPO-THALAMUS

Lateral hypo-thalamus

Central complex

Corticomedial complex

Basolateral complex

Temporal *STF*

AMYGDALA

Hypothalamic NEP

Ventromedial nucleus

Medial forebrain bundle
Optic tract

Superior olivary complex?

Midline raphe glial system

Principal sensory nucleus (V)?
Central trigeminal tract (V)?
Lateral lemniscus?

Subpial GEP

Trapezoid body

Migrating facial motor neurons?

External germinal layer

PONS

RHOMBENCEPHALIC SUPERVENTRICLE

Dorsal rhombic lip

Cerebellar NEP

Pontine NEP

CEREBELLUM (HEMISPHERE)

Ponto-medullary trench

RHOMBENCEPHALIC SUPERVENTRICLE (FUTURE FOURTH VENTRICLE, LATERAL RECESS)

Abducens nucleus (VI)?

Auditory NEP

Upper medullary NEP

Premigratory facial motor neurons?

Ventral rhombic lip

Reticular formation

Cochlear nucleus (dorsal)

Spinal nucleus (V)

Inferior cerebellar peduncle?

MEDULLA

Raphe nuclear complex

Posterior extramural migratory stream
(external cuneate and lateral reticular neurons)

Capsule
Principal nucleus

Inferior olive

CENTRAL CANAL (SPINAL CORD)

Intermediate gray matter

Spinal cord NEP

Dorsal gray matter (substantia gelatinosa)

Subpial GEP

SPINAL CORD

Dorsal funiculus

FONT KEY:
VENTRICULAR DIVISIONS - CAPITALS
Germinal zone - Helvetica bold
Transient structure - Times bold italic
Permanent structure - Times Roman or **Bold**

ABBREVIATIONS:
GEP - Glioepithelium
NEP - Neuroepithelium
SVZ - Subventricular zone

Arrows indicate the presumed *direction of neuron migration* from neuroepithelial sources.

PLATE 25A
CR 31 mm, GW 9.5, C9226
Frontal/Horizontal, Section 504

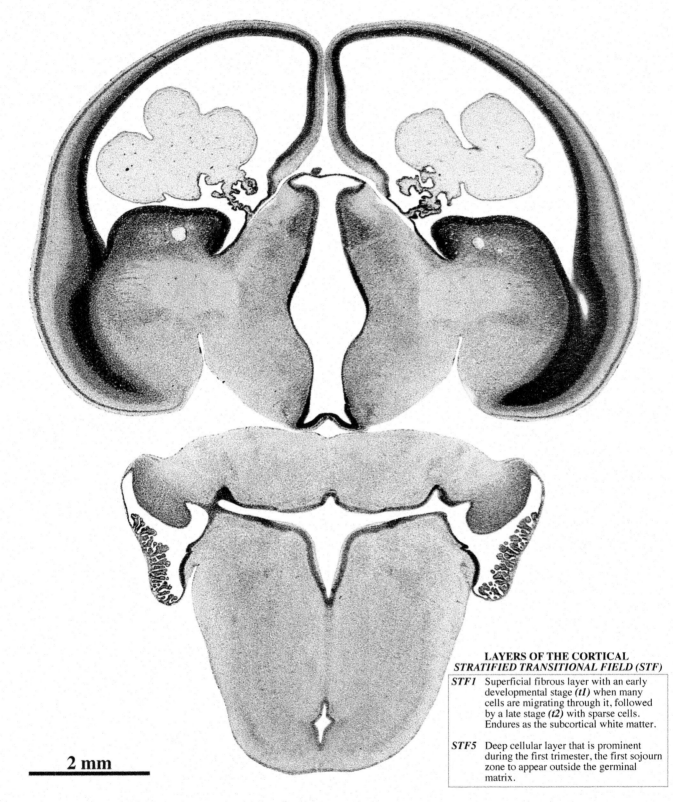

2 mm

LAYERS OF THE CORTICAL
STRATIFIED TRANSITIONAL FIELD (STF)

STF1 Superficial fibrous layer with an early
developmental stage *(t1)* when many
cells are migrating through it, followed
by a late stage *(t2)* with sparse cells.
Endures as the subcortical white matter.

STF5 Deep cellular layer that is prominent
during the first trimester, the first sojourn
zone to appear outside the germinal
matrix.

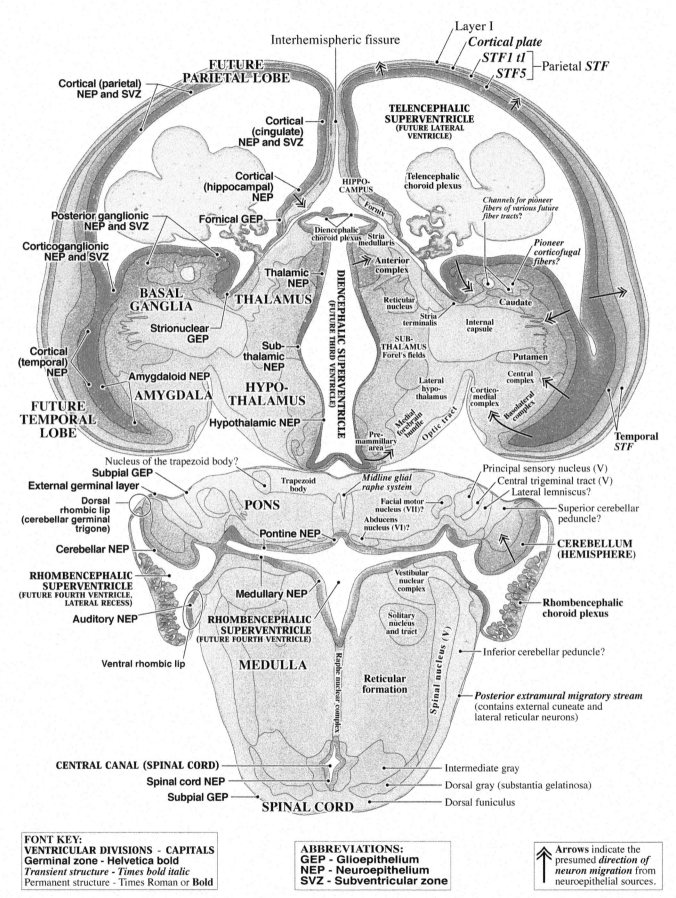

Interhemispheric fissure

Layer I
Cortical plate
STF1 t1
STF5 Parietal *STF*

FUTURE
PARIETAL LOBE

Cortical (parietal)
NEP and SVZ

Cortical
(cingulate)
NEP and SVZ

**TELENCEPHALIC
SUPERVENTRICLE**
(FUTURE LATERAL
VENTRICLE)

Cortical
(hippocampal)
NEP

HIPPO-
CAMPUS

**Telencephalic
choroid plexus**

*Channels for pioneer
fibers of various future
fiber tracts?*

Posterior ganglionic
NEP and SVZ

Fornical GEP

Fornix

*Pioneer
corticofugal
fibers?*

Corticoganglionic
NEP and SVZ

Diencephalic
choroid plexus

*Stria
medullaris*

*Anterior
complex*

Thalamic
NEP

Reticular
nucleus

Caudate

**BASAL
GANGLIA**

THALAMUS

Stria
terminalis

Internal
capsule

Strionuclear
GEP

Sub-
thalamic
NEP

DIENCEPHALIC SUPERVENTRICLE
(FUTURE THIRD VENTRICLE)

**SUB-
THALAMUS**
Forel's fields

Putamen

Cortical
(temporal)
NEP

Amygdaloid NEP

AMYGDALA

**HYPO-
THALAMUS**

*Lateral
hypo-
thalamus*

Central
complex

**FUTURE
TEMPORAL
LOBE**

Hypothalamic NEP

Pre-
mammillary
area

*Medial
forebrain
bundle*

Optic tract

Cortico-
medial
complex

*Basolateral
complex*

Temporal
STF

Nucleus of the trapezoid body?

Subpial GEP

External germinal layer

*Trapezoid
body*

*Midline glial
raphe system*

Principal sensory nucleus (V)
Central trigeminal tract (V)
Lateral lemniscus?

Dorsal
rhombic lip
(cerebellar germinal
trigone)

PONS

Facial motor
nucleus (VII)?

Superior cerebellar
peduncle?

Cerebellar NEP

Abducens
nucleus (VI)?

**RHOMBENCEPHALIC
SUPERVENTRICLE**
(FUTURE FOURTH VENTRICLE,
LATERAL RECESS)

Pontine NEP

**CEREBELLUM
(HEMISPHERE)**

Auditory NEP

Medullary NEP

*Vestibular
nuclear
complex*

**Rhombencephalic
choroid plexus**

**RHOMBENCEPHALIC
SUPERVENTRICLE**
(FUTURE FOURTH VENTRICLE)

*Solitary
nucleus
and tract*

Ventral rhombic lip

MEDULLA

Raphe nuclear complex

**Reticular
formation**

Spinal nucleus (V)

Inferior cerebellar peduncle?

Posterior extramural migratory stream
(contains external cuneate and
lateral reticular neurons)

CENTRAL CANAL (SPINAL CORD)

Spinal cord NEP

Subpial GEP

SPINAL CORD

Intermediate gray

Dorsal gray (substantia gelatinosa)

Dorsal funiculus

FONT KEY:
VENTRICULAR DIVISIONS - CAPITALS
Germinal zone - Helvetica bold
Transient structure - Times bold italic
Permanent structure - Times Roman or **Bold**

ABBREVIATIONS:
GEP - Glioepithelium
NEP - Neuroepithelium
SVZ - Subventricular zone

Arrows indicate the
presumed *direction of
neuron migration* from
neuroepithelial sources.

PLATE 26A
CR 31 mm, GW 9.5, C9226
Frontal/Horizontal, Section 468

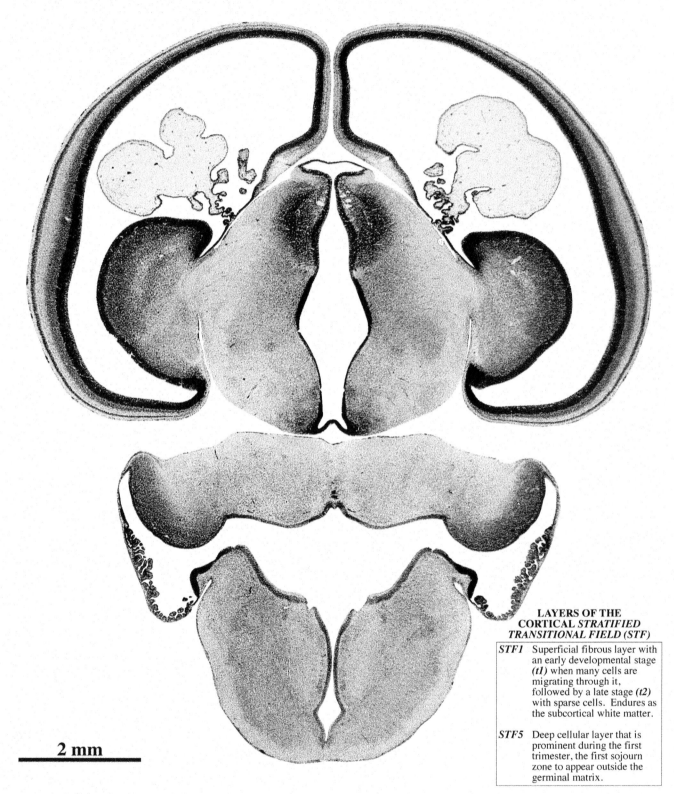

2 mm

LAYERS OF THE
CORTICAL *STRATIFIED*
TRANSITIONAL FIELD (STF)

STF1 Superficial fibrous layer with
an early developmental stage
(t1) when many cells are
migrating through it,
followed by a late stage *(t2)*
with sparse cells. Endures as
the subcortical white matter.

STF5 Deep cellular layer that is
prominent during the first
trimester, the first sojourn
zone to appear outside the
germinal matrix.

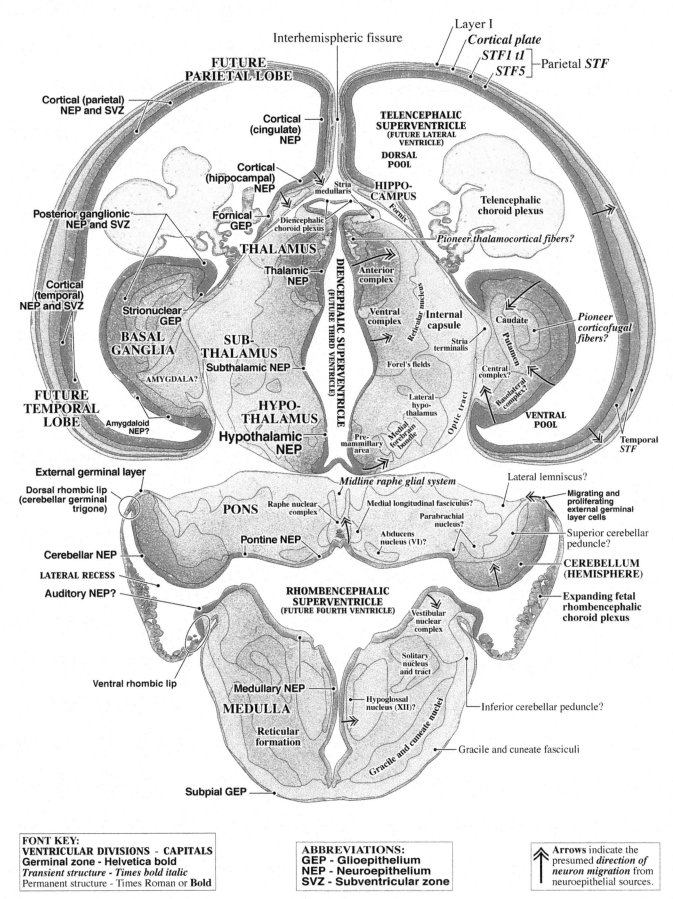

Layer I
Cortical plate
STF1 t1
STF5 — Parietal *STF*

Interhemispheric fissure

FUTURE
PARIETAL LOBE

Cortical (parietal)
NEP and SVZ

Cortical
(cingulate)
NEP

TELENCEPHALIC
SUPERVENTRICLE
(FUTURE LATERAL
VENTRICLE)

DORSAL
POOL

Cortical
(hippocampal)
NEP

Stria
medullaris

HIPPO-
CAMPUS

Telencephalic
choroid plexus

Posterior ganglionic
NEP and SVZ

Fornical
GEP

Diencephalic
choroid plexus

Fornix

Pioneer thalamocortical fibers?

THALAMUS

Thalamic
NEP

Anterior
complex

Cortical
(temporal)
NEP and SVZ

Strionuclear
GEP

Ventral
complex

Internal
capsule

Reticular nucleus

Caudate

*Pioneer
corticofugal
fibers?*

Putamen

BASAL
GANGLIA

SUB-
THALAMUS

Subthalamic NEP

Stria
terminalis

Central
complex?

AMYGDALA?

Forel's fields

Basolateral
complex?

FUTURE
TEMPORAL
LOBE

HYPO-
THALAMUS

Lateral
hypo-
thalamus

VENTRAL
POOL

Amygdaloid
NEP?

Hypothalamic
NEP

Pre-
mammillary
area

Medial
forebrain
bundle

Optic tract

Temporal
STF

DIENCEPHALIC SUPERVENTRICLE (FUTURE THIRD VENTRICLE)

External germinal layer

Midline raphe glial system

Lateral lemniscus?

Dorsal rhombic lip
(cerebellar germinal
trigone)

PONS

Raphe nuclear
complex

Medial longitudinal fasciculus?

Migrating and
proliferating
external germinal
layer cells

Cerebellar NEP

Pontine NEP

Parabrachial
nucleus?

Abducens
nucleus (VI)?

Superior cerebellar
peduncle?

LATERAL RECESS

CEREBELLUM
(HEMISPHERE)

Auditory NEP?

RHOMBENCEPHALIC
SUPERVENTRICLE
(FUTURE FOURTH VENTRICLE)

Vestibular
nuclear
complex

Expanding fetal
rhombencephalic
choroid plexus

Ventral rhombic lip

Solitary
nucleus
and tract

Medullary NEP

Hypoglossal
nucleus (XII)?

Inferior cerebellar peduncle?

MEDULLA

Gracile and cuneate nuclei

Reticular
formation

Gracile and cuneate fasciculi

Subpial GEP

FONT KEY:
VENTRICULAR DIVISIONS - CAPITALS
Germinal zone - Helvetica bold
Transient structure - Times bold italic
Permanent structure - Times Roman or **Bold**

ABBREVIATIONS:
GEP - Glioepithelium
NEP - Neuroepithelium
SVZ - Subventricular zone

Arrows indicate the
presumed *direction of
neuron migration* from
neuroepithelial sources.

PLATE 27A
CR 31 mm, GW 9.5, C9226
Frontal/Horizontal, Section 438

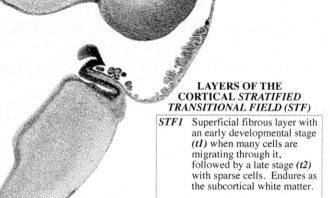

2 mm

LAYERS OF THE
CORTICAL *STRATIFIED*
TRANSITIONAL FIELD (STF)

STF1	Superficial fibrous layer with an early developmental stage *(t1)* when many cells are migrating through it, followed by a late stage *(t2)* with sparse cells. Endures as the subcortical white matter.
STF5	Deep cellular layer that is prominent during the first trimester, the first sojourn zone to appear outside the germinal matrix.

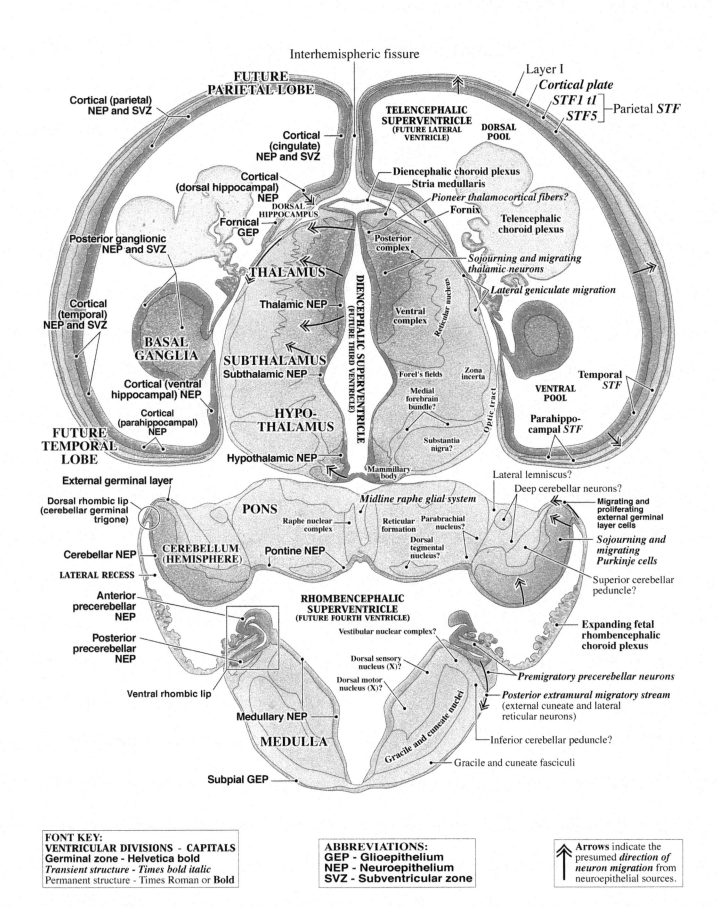

Interhemispheric fissure

FUTURE
PARIETAL LOBE

Cortical (parietal)
NEP and SVZ

Cortical
(cingulate)
NEP and SVZ

Cortical
(dorsal hippocampal)
NEP

Fornical
GEP

DORSAL
HIPPOCAMPUS

THALAMUS

Posterior ganglionic
NEP and SVZ

Cortical
(temporal)
NEP and SVZ

BASAL
GANGLIA

Thalamic NEP

SUBTHALAMUS
Subthalamic NEP

Cortical (ventral
hippocampal) NEP

Cortical
(parahippocampal)
NEP

HYPO-
THALAMUS

FUTURE
TEMPORAL
LOBE

Hypothalamic NEP

External germinal layer

Dorsal rhombic lip
(cerebellar germinal
trigone)

PONS

Cerebellar NEP

LATERAL RECESS

CEREBELLUM
(HEMISPHERE)

Anterior
precerebellar
NEP

Posterior
precerebellar
NEP

Raphe nuclear
complex

Pontine NEP

Ventral rhombic lip

RHOMBENCEPHALIC
SUPERVENTRICLE
(FUTURE FOURTH VENTRICLE)

Vestibular nuclear complex?

Dorsal sensory
nucleus (X)?

Dorsal motor
nucleus (X)?

Medullary NEP

MEDULLA

Subpial GEP

Layer I
Cortical plate
STF1 t1
STF5

Parietal STF

TELENCEPHALIC
SUPERVENTRICLE
(FUTURE LATERAL
VENTRICLE)

DORSAL
POOL

Diencephalic choroid plexus
Stria medullaris
Pioneer thalamocortical fibers?
Fornix

Posterior
complex

Telencephalic
choroid plexus

Sojourning and migrating
thalamic neurons

Lateral geniculate migration

Ventral
complex

Reticular nucleus

DIENCEPHALIC SUPERVENTRICLE
(FUTURE THIRD VENTRICLE)

Forel's fields

Zona
incerta

Medial
forebrain
bundle?

Optic tract

Temporal
STF

VENTRAL
POOL

Parahippo-
campal STF

Substantia
nigra?

Mammillary
body

Lateral lemniscus?

Deep cerebellar neurons?

Midline raphe glial system

Reticular
formation

Parabrachial
nucleus?

Dorsal
tegmental
nucleus?

Migrating and
proliferating
external germinal
layer cells

Sojourning and
migrating
Purkinje cells

Superior cerebellar
peduncle?

Expanding fetal
rhombencephalic
choroid plexus

Premigratory precerebellar neurons

Posterior extramural migratory stream
(external cuneate and lateral
reticular neurons)

Inferior cerebellar peduncle?

Gracile and cuneate nuclei

Gracile and cuneate fasciculi

FONT KEY:
VENTRICULAR DIVISIONS - CAPITALS
Germinal zone - Helvetica bold
Transient structure - Times bold italic
Permanent structure - Times Roman or **Bold**

ABBREVIATIONS:
GEP - Glioepithelium
NEP - Neuroepithelium
SVZ - Subventricular zone

Arrows indicate the
presumed *direction of
neuron migration* from
neuroepithelial sources.

PLATE 28A
CR 31 mm, GW 9.5, C9226
Frontal/Horizontal, Section 402

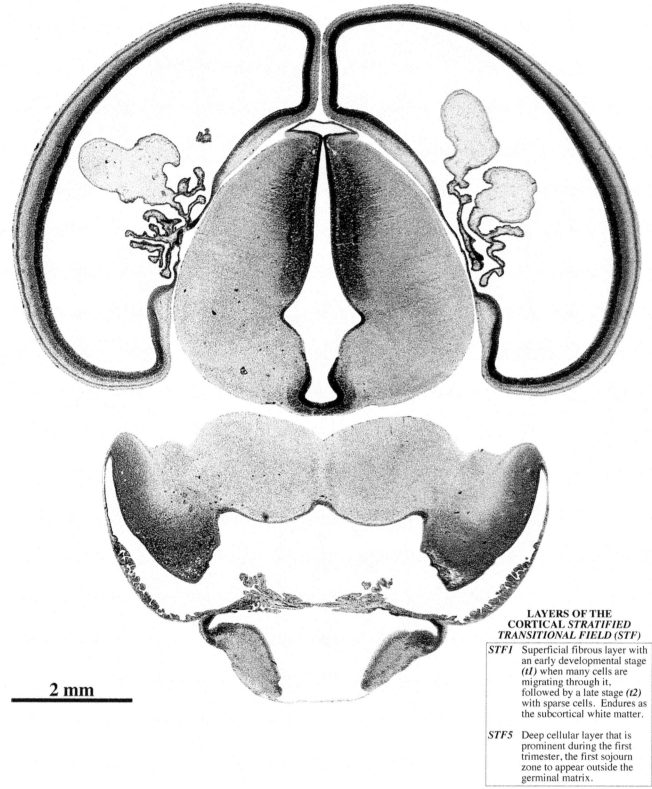

2 mm

LAYERS OF THE
CORTICAL *STRATIFIED*
TRANSITIONAL FIELD (STF)

STF1 Superficial fibrous layer with
an early developmental stage
(t1) when many cells are
migrating through it,
followed by a late stage *(t2)*
with sparse cells. Endures as
the subcortical white matter.

STF5 Deep cellular layer that is
prominent during the first
trimester, the first sojourn
zone to appear outside the
germinal matrix.

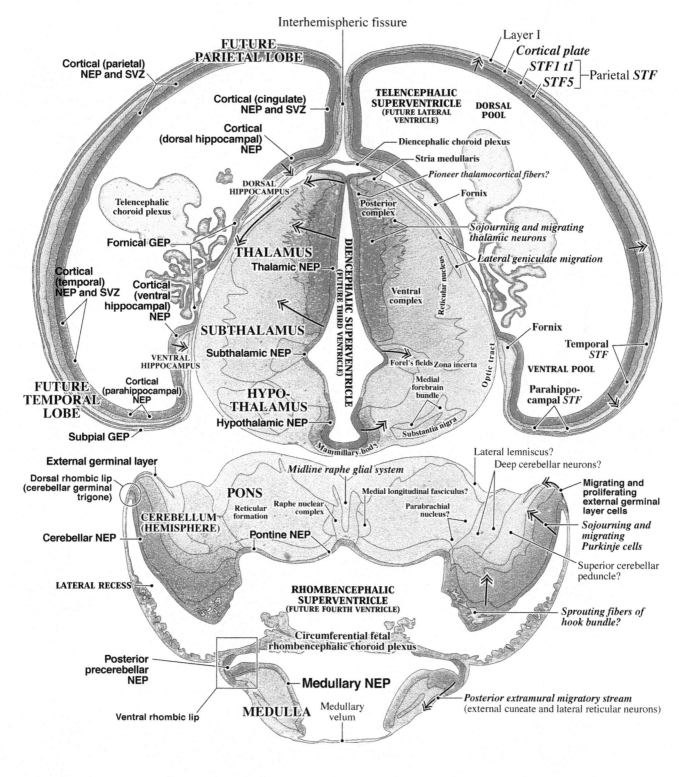

Interhemispheric fissure

Layer I
Cortical plate
STF1 t1
STF5 ⎤ Parietal *STF*

Cortical (parietal)
NEP and SVZ

**FUTURE
PARIETAL LOBE**

Cortical (cingulate)
NEP and SVZ

**TELENCEPHALIC
SUPERVENTRICLE**
(FUTURE LATERAL
VENTRICLE)

**DORSAL
POOL**

Cortical
(dorsal hippocampal)
NEP

Diencephalic choroid plexus

Stria medullaris

Pioneer thalamocortical fibers?

Telencephalic
choroid plexus

*DORSAL
HIPPOCAMPUS*

Posterior
complex

Fornix

*Sojourning and migrating
thalamic neurons*

Fornical GEP

THALAMUS

DIENCEPHALIC SUPERVENTRICLE
(FUTURE THIRD VENTRICLE)

Lateral geniculate migration

Cortical
(temporal)
NEP and SVZ

Thalamic NEP

Reticular nucleus

Cortical
(ventral
hippocampal)
NEP

Ventral
complex

SUBTHALAMUS

Subthalamic NEP

Fornix

**Temporal
*STF***

*VENTRAL
HIPPOCAMPUS*

Forel's fields Zona incerta

Optic tract

VENTRAL POOL

**FUTURE
TEMPORAL
LOBE**

Cortical
(parahippocampal)
NEP

**HYPO-
THALAMUS**

Medial
forebrain
bundle

Parahippo-
campal *STF*

Hypothalamic NEP

Subpial GEP

Substantia nigra

Mammillary bodies

External germinal layer

Midline raphe glial system

Lateral lemniscus?
Deep cerebellar neurons?

Dorsal rhombic lip
(cerebellar germinal
trigone)

PONS

Medial longitudinal fasciculus?

**Migrating and
proliferating
external germinal
layer cells**

Reticular
formation

Raphe nuclear
complex

Parabrachial
nucleus?

*Sojourning and
migrating
Purkinje cells*

**CEREBELLUM
(HEMISPHERE)**

Cerebellar NEP

Pontine NEP

Superior cerebellar
peduncle?

LATERAL RECESS

**RHOMBENCEPHALIC
SUPERVENTRICLE**
(FUTURE FOURTH VENTRICLE)

*Sprouting fibers of
hook bundle?*

*Circumferential fetal
rhombencephalic choroid plexus*

Posterior
precerebellar
NEP

Medullary NEP

Ventral rhombic lip

MEDULLA

Medullary
velum

Posterior extramural migratory stream
(external cuneate and lateral reticular neurons)

FONT KEY:
VENTRICULAR DIVISIONS - CAPITALS
Germinal zone - Helvetica bold
Transient structure - Times bold italic
Permanent structure - Times Roman or **Bold**

ABBREVIATIONS:
GEP - Glioepithelium
NEP - Neuroepithelium

Arrows indicate the
presumed *direction of
neuron migration* from
neuroepithelial sources.

PLATE 29A
CR 31 mm, GW 9.5, C9226
Frontal/Horizontal, Section 390

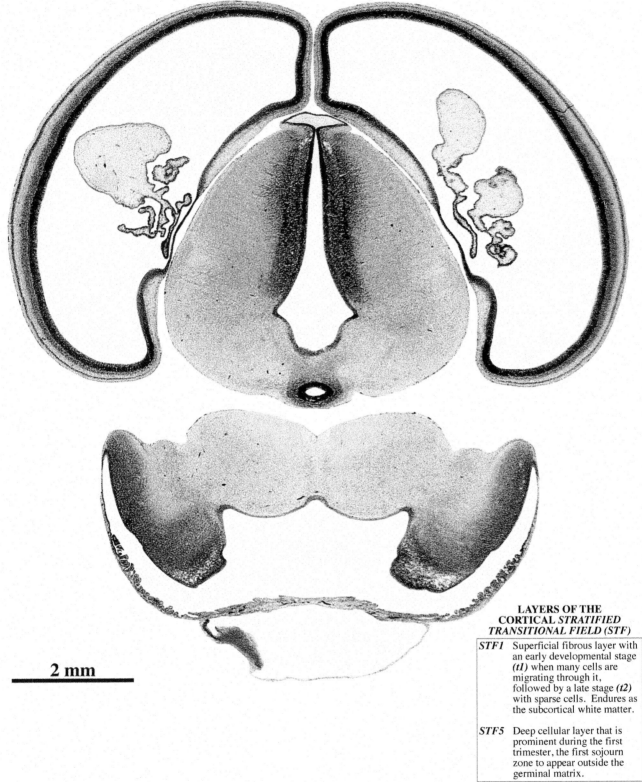

2 mm

LAYERS OF THE
CORTICAL *STRATIFIED*
TRANSITIONAL FIELD (STF)

STF1 Superficial fibrous layer with
an early developmental stage
(t1) when many cells are
migrating through it,
followed by a late stage *(t2)*
with sparse cells. Endures as
the subcortical white matter.

STF5 Deep cellular layer that is
prominent during the first
trimester, the first sojourn
zone to appear outside the
germinal matrix.

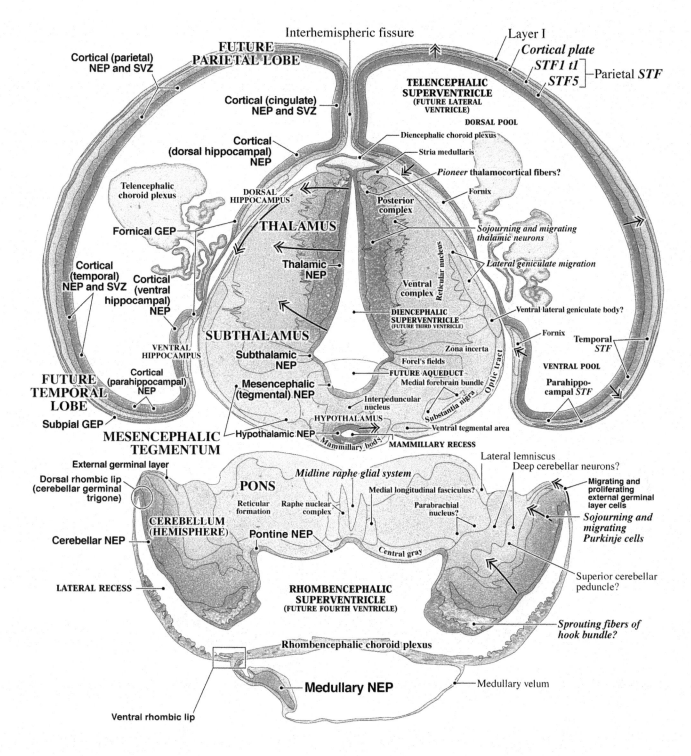

Interhemispheric fissure

Layer I
Cortical plate
STF1 t1 ⎫ Parietal *STF*
STF5 ⎭

Cortical (parietal) NEP and SVZ

FUTURE PARIETAL LOBE

Cortical (cingulate) NEP and SVZ

TELENCEPHALIC SUPERVENTRICLE (FUTURE LATERAL VENTRICLE)

DORSAL POOL

Cortical (dorsal hippocampal) NEP

Diencephalic choroid plexus

Stria medullaris

Pioneer thalamocortical fibers?

Telencephalic choroid plexus

DORSAL HIPPOCAMPUS

Posterior complex

Fornix

Fornical GEP

THALAMUS

Sojourning and migrating thalamic neurons

Lateral geniculate migration

Cortical (temporal) NEP and SVZ

Cortical (ventral hippocampal) NEP

Thalamic NEP

Ventral complex

Reticular nucleus

Ventral lateral geniculate body?

VENTRAL HIPPOCAMPUS

SUBTHALAMUS

DIENCEPHALIC SUPERVENTRICLE (FUTURE THIRD VENTRICLE)

Fornix

Temporal *STF*

Cortical (parahippocampal) NEP

Subthalamic NEP

Zona incerta

Forel's fields

FUTURE AQUEDUCT

Optic tract

VENTRAL POOL

FUTURE TEMPORAL LOBE

Mesencephalic (tegmental) NEP

Medial forebrain bundle

Parahippocampal *STF*

Subpial GEP

Hypothalamic NEP

Interpeduncular nucleus

Substantia nigra

MESENCEPHALIC TEGMENTUM

HYPOTHALAMUS

Ventral tegmental area

Mammillary body

MAMMILLARY RECESS

External germinal layer

Lateral lemniscus
Deep cerebellar neurons?

Dorsal rhombic lip (cerebellar germinal trigone)

Midline raphe glial system

PONS

Medial longitudinal fasciculus?

Migrating and proliferating external germinal layer cells

Reticular formation

Raphe nuclear complex

Parabrachial nucleus?

Sojourning and migrating Purkinje cells

CEREBELLUM (HEMISPHERE)

Pontine NEP

Cerebellar NEP

Central gray

Superior cerebellar peduncle?

LATERAL RECESS

RHOMBENCEPHALIC SUPERVENTRICLE (FUTURE FOURTH VENTRICLE)

Sprouting fibers of hook bundle?

Rhombencephalic choroid plexus

Medullary NEP

Medullary velum

Ventral rhombic lip

FONT KEY:
VENTRICULAR DIVISIONS - CAPITALS
Germinal zone - Helvetica bold
Transient structure - Times bold italic
Permanent structure - Times Roman or **Bold**

ABBREVIATIONS:
GEP - Glioepithelium
NEP - Neuroepithelium

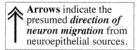

Arrows indicate the presumed *direction of neuron migration* from neuroepithelial sources.

PLATE 30A
CR 31 mm, GW 9.5, C9226
Frontal/Horizontal, Section 324

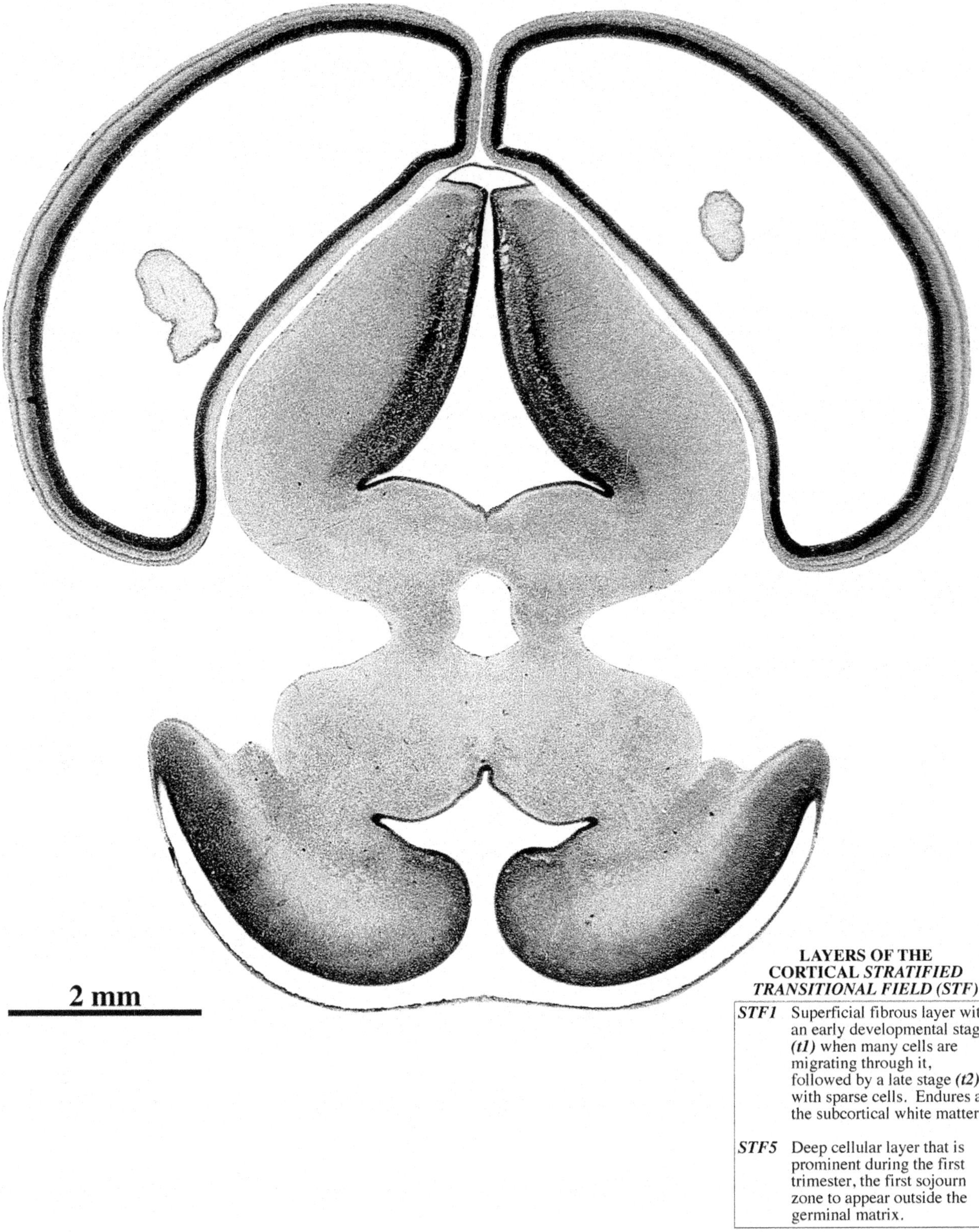

2 mm

LAYERS OF THE
CORTICAL *STRATIFIED*
TRANSITIONAL FIELD (STF)

STF1 Superficial fibrous layer with
an early developmental stage
(t1) when many cells are
migrating through it,
followed by a late stage *(t2)*
with sparse cells. Endures as
the subcortical white matter.

STF5 Deep cellular layer that is
prominent during the first
trimester, the first sojourn
zone to appear outside the
germinal matrix.

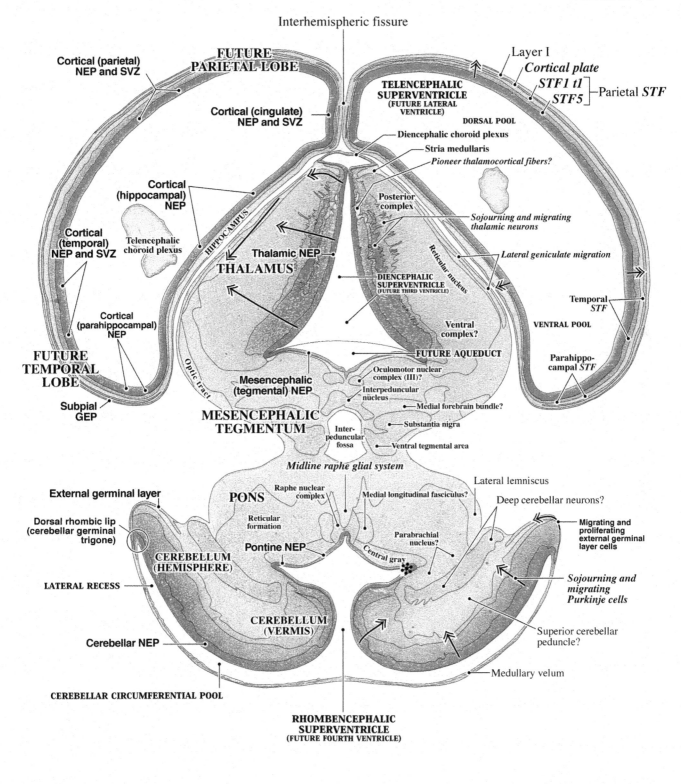

Interhemispheric fissure

Cortical (parietal) NEP and SVZ

FUTURE PARIETAL LOBE

Layer I
Cortical plate
STF1 t1
STF5 — Parietal *STF*

Cortical (cingulate) NEP and SVZ

TELENCEPHALIC SUPERVENTRICLE (FUTURE LATERAL VENTRICLE)

DORSAL POOL

Diencephalic choroid plexus

Stria medullaris

Pioneer thalamocortical fibers?

Cortical (hippocampal) NEP

Posterior complex

Sojourning and migrating thalamic neurons

Cortical (temporal) NEP and SVZ

Telencephalic choroid plexus

Lateral geniculate migration

HIPPOCAMPUS

Thalamic NEP

THALAMUS

Reticular nucleus

DIENCEPHALIC SUPERVENTRICLE (FUTURE THIRD VENTRICLE)

Temporal *STF*

Cortical (parahippocampal) NEP

FUTURE TEMPORAL LOBE

Ventral complex?

VENTRAL POOL

FUTURE AQUEDUCT

Parahippocampal *STF*

Optic tract

Mesencephalic (tegmental) NEP

Oculomotor nuclear complex (III)?

Subpial GEP

Interpeduncular nucleus

Medial forebrain bundle?

MESENCEPHALIC TEGMENTUM

Interpeduncular fossa

Substantia nigra

Ventral tegmental area

Midline raphe glial system

Lateral lemniscus

External germinal layer

PONS

Raphe nuclear complex

Medial longitudinal fasciculus?

Deep cerebellar neurons?

Dorsal rhombic lip (cerebellar germinal trigone)

Reticular formation

Parabrachial nucleus?

Migrating and proliferating external germinal layer cells

CEREBELLUM (HEMISPHERE)

Pontine NEP

Central gray

Sojourning and migrating Purkinje cells

LATERAL RECESS

CEREBELLUM (VERMIS)

Superior cerebellar peduncle?

Cerebellar NEP

Medullary velum

CEREBELLAR CIRCUMFERENTIAL POOL

RHOMBENCEPHALIC SUPERVENTRICLE (FUTURE FOURTH VENTRICLE)

FONT KEY:
VENTRICULAR DIVISIONS - CAPITALS
Germinal zone - Helvetica bold
Transient structure - Times bold italic
Permanent structure - Times Roman or **Bold**

ABBREVIATIONS:
GEP - Glioepithelium
NEP - Neuroepithelium

Arrows indicate the presumed *direction of neuron migration* from neuroepithelial sources.

PLATE 31A
CR 31 mm, GW 9.5, C9226
Frontal/Horizontal, Section 276

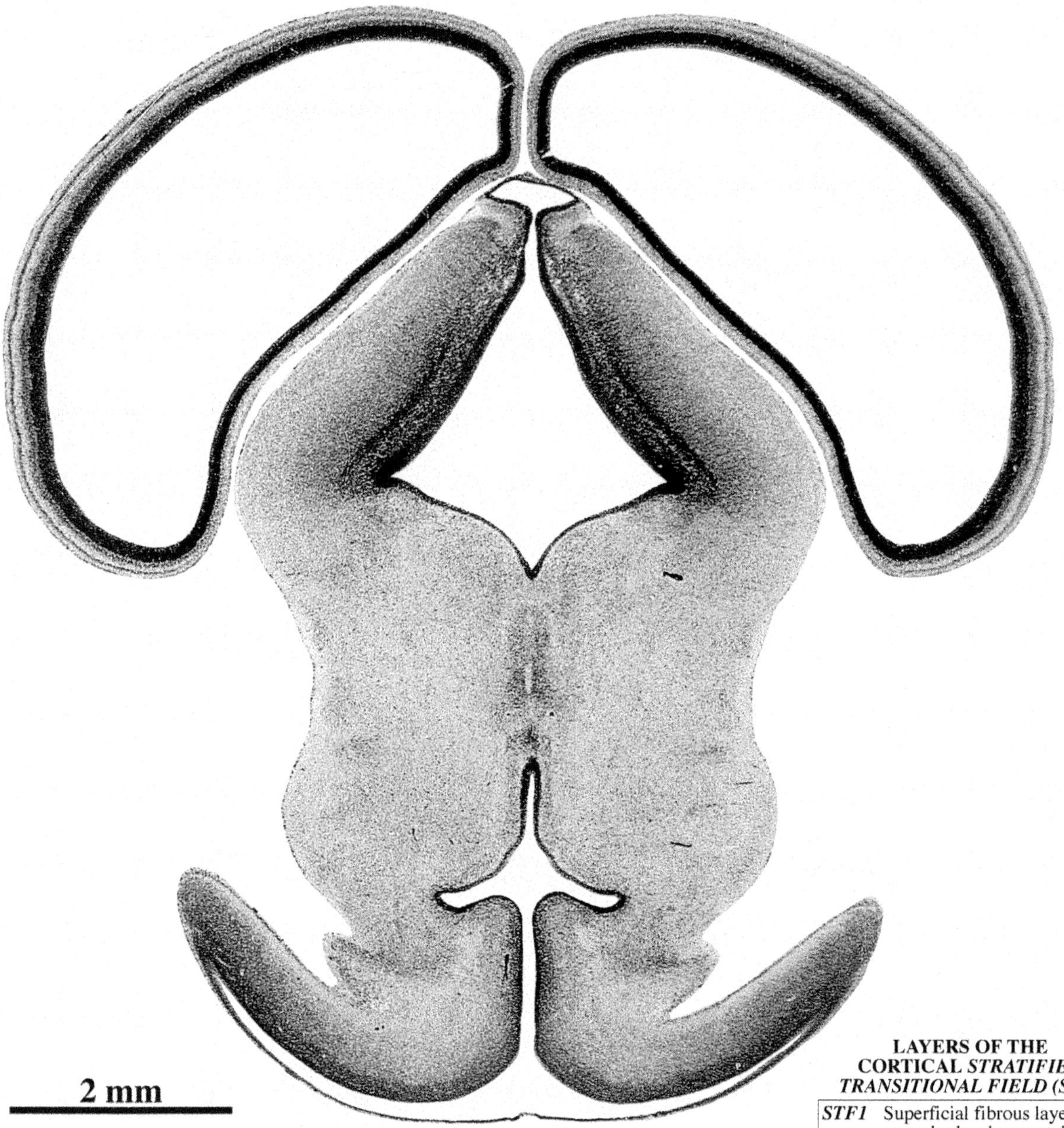

2 mm

LAYERS OF THE
CORTICAL *STRATIFIED*
TRANSITIONAL FIELD (STF)

STF1 Superficial fibrous layer with
an early developmental stage
(t1) when many cells are
migrating through it,
followed by a late stage *(t2)*
with sparse cells. Endures as
the subcortical white matter.

STF5 Deep cellular layer that is
prominent during the first
trimester, the first sojourn
zone to appear outside the
germinal matrix.

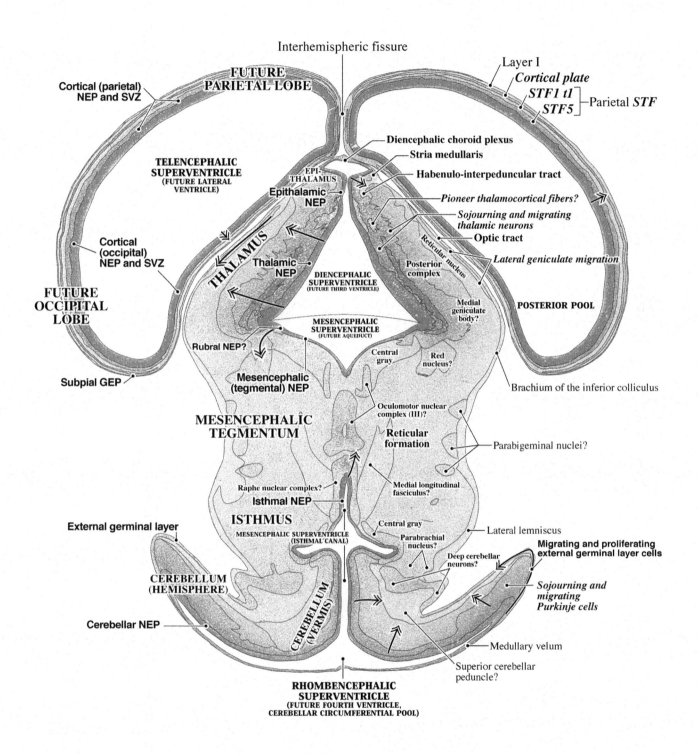

Interhemispheric fissure

FUTURE
PARIETAL LOBE

Cortical (parietal)
NEP and SVZ

Layer I
Cortical plate
STF1 t1
STF5 }Parietal *STF*

TELENCEPHALIC
SUPERVENTRICLE
(FUTURE LATERAL
VENTRICLE)

Diencephalic choroid plexus
Stria medullaris
EPI-
THALAMUS
Habenulo-interpeduncular tract
Epithalamic
NEP
Pioneer thalamocortical fibers?
*Sojourning and migrating
thalamic neurons*
Optic tract
Reticular nucleus
Lateral geniculate migration

THALAMUS

Thalamic
NEP

Cortical
(occipital)
NEP and SVZ

DIENCEPHALIC
SUPERVENTRICLE
(FUTURE THIRD VENTRICLE)

Posterior
complex

FUTURE
OCCIPITAL
LOBE

Medial
geniculate
body?

POSTERIOR POOL

MESENCEPHALIC
SUPERVENTRICLE
(FUTURE AQUEDUCT)

Rubral NEP?

Central
gray

Red
nucleus?

Subpial GEP

Mesencephalic
(tegmental) NEP

Brachium of the inferior colliculus

Oculomotor nuclear
complex (III)?

MESENCEPHALIC
TEGMENTUM

**Reticular
formation**

Parabigeminal nuclei?

Raphe nuclear complex?

Medial longitudinal
fasciculus?

Isthmal NEP

ISTHMUS

Central gray

Lateral lemniscus

External germinal layer

MESENCEPHALIC SUPERVENTRICLE
(ISTHMAL CANAL)

Parabrachial
nucleus?

**Migrating and proliferating
external germinal layer cells**

CEREBELLUM
(HEMISPHERE)

Deep cerebellar
neurons?

*Sojourning and
migrating
Purkinje cells*

Cerebellar NEP

CEREBELLUM
(VERMIS)

Medullary velum

Superior cerebellar
peduncle?

RHOMBENCEPHALIC
SUPERVENTRICLE
(FUTURE FOURTH VENTRICLE,
CEREBELLAR CIRCUMFERENTIAL POOL)

FONT KEY:
VENTRICULAR DIVISIONS - CAPITALS
Germinal zone - Helvetica bold
Transient structure - Times bold italic
Permanent structure - Times Roman or **Bold**

ABBREVIATIONS:
GEP - Glioepithelium
NEP - Neuroepithelium

Arrows indicate the
presumed *direction of
neuron migration* from
neuroepithelial sources.

PLATE 32A
CR 31 mm, GW 9.5, C9226
Frontal/Horizontal, Section 252

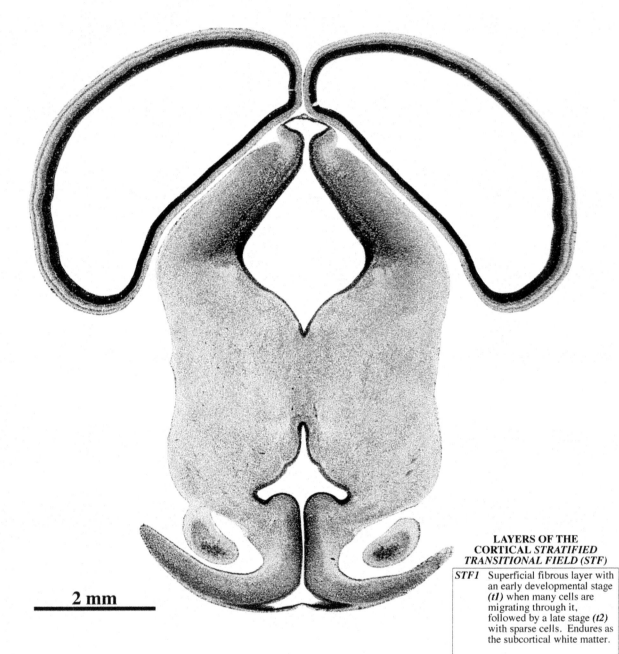

2 mm

LAYERS OF THE
CORTICAL *STRATIFIED*
TRANSITIONAL FIELD (STF)

STF1 Superficial fibrous layer with
an early developmental stage
(t1) when many cells are
migrating through it,
followed by a late stage *(t2)*
with sparse cells. Endures as
the subcortical white matter.

STF5 Deep cellular layer that is
prominent during the first
trimester, the first sojourn
zone to appear outside the
germinal matrix.

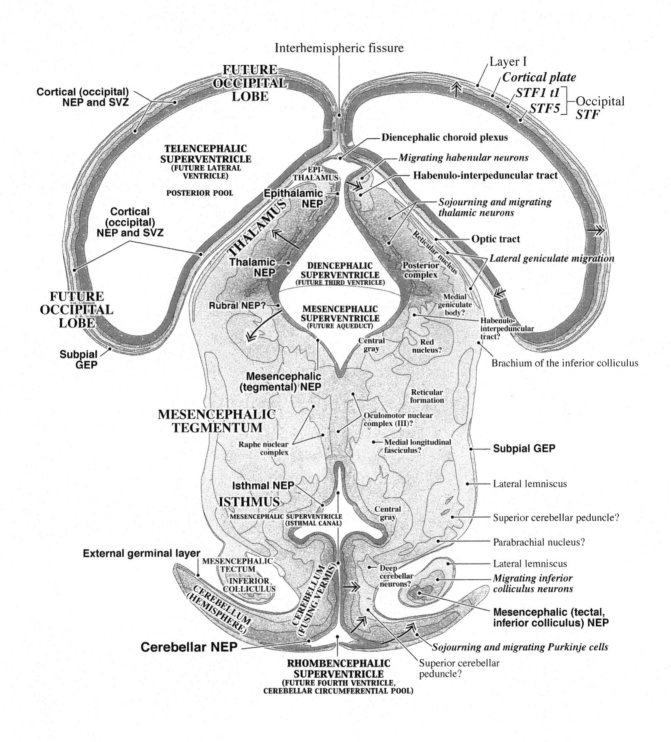

Interhemispheric fissure

FUTURE OCCIPITAL LOBE

Cortical (occipital) NEP and SVZ

Layer I
Cortical plate
STF1 t1
STF5
Occipital *STF*

Diencephalic choroid plexus

Migrating habenular neurons

Habenulo-interpeduncular tract

TELENCEPHALIC SUPERVENTRICLE (FUTURE LATERAL VENTRICLE)

POSTERIOR POOL

EPI-THAL'AMUS

Epithalamic NEP

Sojourning and migrating thalamic neurons

THAL'AMUS

Cortical (occipital) NEP and SVZ

Thalamic NEP

DIENCEPHALIC SUPERVENTRICLE (FUTURE THIRD VENTRICLE)

Reticular nucleus

Posterior complex

Optic tract

Lateral geniculate migration

FUTURE OCCIPITAL LOBE

Rubral NEP?

MESENCEPHALIC SUPERVENTRICLE (FUTURE AQUEDUCT)

Central gray

Red nucleus?

Medial geniculate body?

Habenulo-interpeduncular tract?

Subpial GEP

Brachium of the inferior colliculus

Mesencephalic (tegmental) NEP

Reticular formation

Oculomotor nuclear complex (III)?

MESENCEPHALIC TEGMENTUM

Raphe nuclear complex

Medial longitudinal fasciculus?

Subpial GEP

Lateral lemniscus

Isthmal NEP

ISTHMUS

MESENCEPHALIC SUPERVENTRICLE (ISTHMAL CANAL)

Central gray

Superior cerebellar peduncle?

Parabrachial nucleus?

External germinal layer

MESENCEPHALIC TECTUM

INFERIOR COLLICULUS

CEREBELLUM (HEMISPHERE)

CEREBELLUM (FUSING VERMIS)

Deep cerebellar neurons?

Lateral lemniscus

Migrating inferior colliculus neurons

Mesencephalic (tectal, inferior colliculus) NEP

Cerebellar NEP

RHOMBENCEPHALIC SUPERVENTRICLE (FUTURE FOURTH VENTRICLE, CEREBELLAR CIRCUMFERENTIAL POOL)

Superior cerebellar peduncle?

Sojourning and migrating Purkinje cells

FONT KEY:
VENTRICULAR DIVISIONS - CAPITALS
Germinal zone - Helvetica bold
Transient structure - Times bold italic
Permanent structure - Times Roman or **Bold**

ABBREVIATIONS:
GEP - Glioepithelium
NEP - Neuroepithelium

Arrows indicate the presumed *direction of neuron migration* from neuroepithelial sources.

PLATE 33A
CR 31 mm, GW 9.5, C9226
Frontal/Horizontal, Section 234

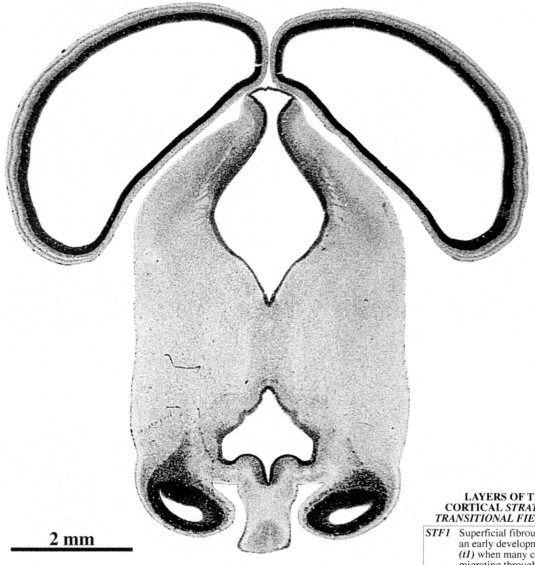

2 mm

LAYERS OF THE
CORTICAL *STRATIFIED*
***TRANSITIONAL FIELD* (STF)**

STF1 Superficial fibrous layer with
an early developmental stage
(t1) when many cells are
migrating through it,
followed by a late stage *(t2)*
with sparse cells. Endures as
the subcortical white matter.

STF5 Deep cellular layer that is
prominent during the first
trimester, the first sojourn
zone to appear outside the
germinal matrix.

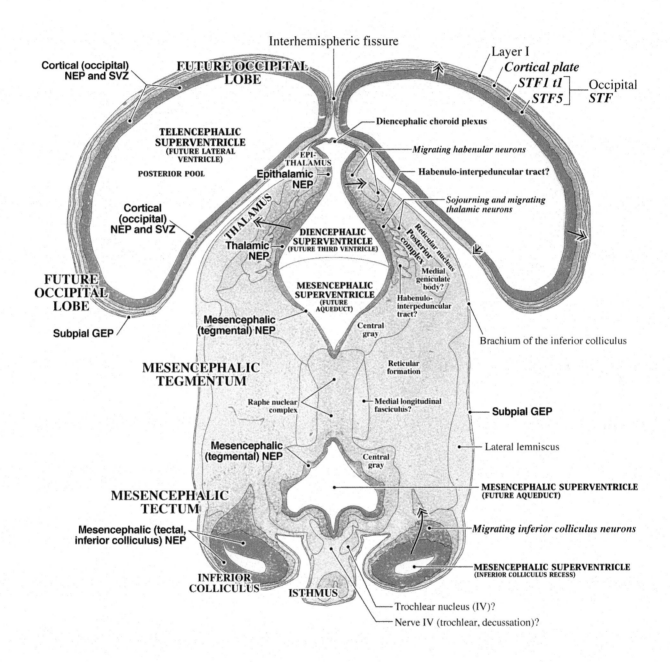

Interhemispheric fissure

Cortical (occipital) NEP and SVZ

FUTURE OCCIPITAL LOBE

Layer I

Cortical plate

STF1 tl

STF5

Occipital *STF*

Diencephalic choroid plexus

TELENCEPHALIC SUPERVENTRICLE (FUTURE LATERAL VENTRICLE)

POSTERIOR POOL

EPI-THALAMUS

Epithalamic NEP

Migrating habenular neurons

Habenulo-interpeduncular tract?

Cortical (occipital) NEP and SVZ

THALAMUS

Sojourning and migrating thalamic neurons

Thalamic NEP

DIENCEPHALIC SUPERVENTRICLE (FUTURE THIRD VENTRICLE)

Reticular nucleus

Posterior complex

FUTURE OCCIPITAL LOBE

MESENCEPHALIC SUPERVENTRICLE (FUTURE AQUEDUCT)

Medial geniculate body?

Subpial GEP

Mesencephalic (tegmental) NEP

Central gray

Habenulo-interpeduncular tract?

Brachium of the inferior colliculus

MESENCEPHALIC TEGMENTUM

Reticular formation

Raphe nuclear complex

Medial longitudinal fasciculus?

Subpial GEP

Mesencephalic (tegmental) NEP

Central gray

Lateral lemniscus

MESENCEPHALIC SUPERVENTRICLE (FUTURE AQUEDUCT)

MESENCEPHALIC TECTUM

Migrating inferior colliculus neurons

Mesencephalic (tectal, inferior colliculus) NEP

MESENCEPHALIC SUPERVENTRICLE (INFERIOR COLLICULUS RECESS)

INFERIOR COLLICULUS

ISTHMUS

Trochlear nucleus (IV)?

Nerve IV (trochlear, decussation)?

FONT KEY:
VENTRICULAR DIVISIONS - CAPITALS
Germinal zone - Helvetica bold
Transient structure - Times bold italic
Permanent structure - Times Roman or **Bold**

ABBREVIATIONS:
GEP - Glioepithelium
NEP - Neuroepithelium

Arrows indicate the presumed *direction of neuron migration* from neuroepithelial sources.

PLATE 34A
CR 31 mm, GW 9.5, C9226
Frontal/Horizontal, Section 206

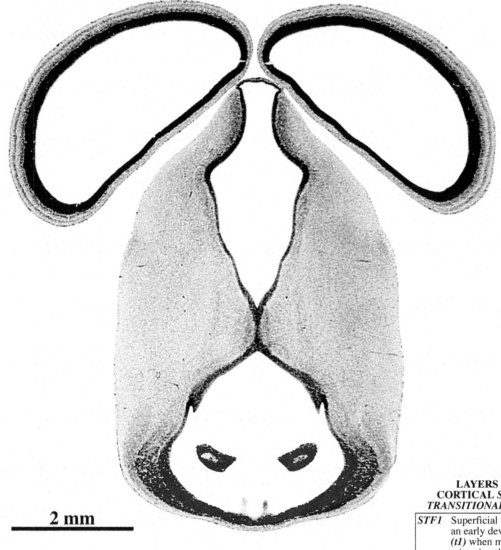

2 mm

LAYERS OF THE
CORTICAL *STRATIFIED*
TRANSITIONAL FIELD (STF)

STF1 Superficial fibrous layer with
an early developmental stage
(t1) when many cells are
migrating through it,
followed by a late stage *(t2)*
with sparse cells. Endures as
the subcortical white matter.

STF5 Deep cellular layer that is
prominent during the first
trimester, the first sojourn
zone to appear outside the
germinal matrix.

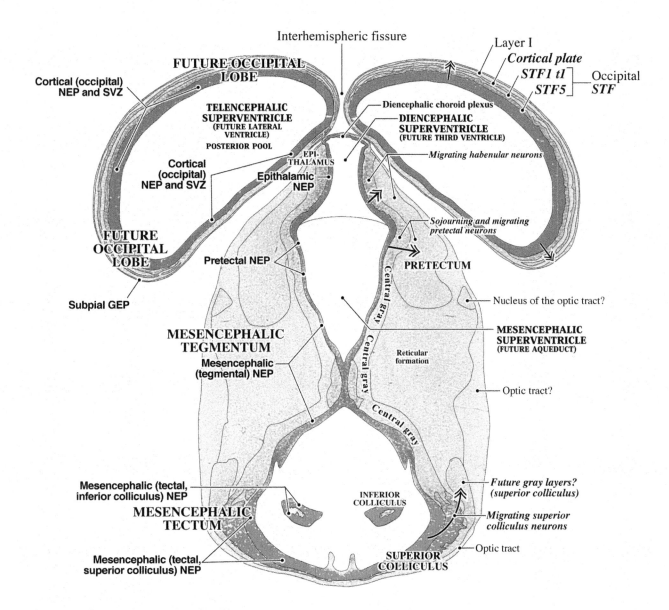

Interhemispheric fissure

Layer I

Cortical plate

STF1 t1

STF5

Occipital *STF*

FUTURE OCCIPITAL LOBE

Cortical (occipital) NEP and SVZ

Diencephalic choroid plexus

DIENCEPHALIC SUPERVENTRICLE (FUTURE THIRD VENTRICLE)

TELENCEPHALIC SUPERVENTRICLE (FUTURE LATERAL VENTRICLE)

POSTERIOR POOL

Migrating habenular neurons

EPI-THALAMUS

Cortical (occipital) NEP and SVZ

Epithalamic NEP

FUTURE OCCIPITAL LOBE

Pretectal NEP

Sojourning and migrating pretectal neurons

PRETECTUM

Nucleus of the optic tract?

Subpial GEP

Central gray

Central gray

MESENCEPHALIC TEGMENTUM

Mesencephalic (tegmental) NEP

Reticular formation

MESENCEPHALIC SUPERVENTRICLE (FUTURE AQUEDUCT)

Optic tract?

Central gray

Mesencephalic (tectal, inferior colliculus) NEP

INFERIOR COLLICULUS

Future gray layers? *(superior colliculus)*

MESENCEPHALIC TECTUM

Migrating superior colliculus neurons

Optic tract

Mesencephalic (tectal, superior colliculus) NEP

SUPERIOR COLLICULUS

FONT KEY:
VENTRICULAR DIVISIONS - CAPITALS
Germinal zone - Helvetica bold
Transient structure - Times bold italic
Permanent structure - Times Roman or **Bold**

ABBREVIATIONS:
GEP - Glioepithelium
NEP - Neuroepithelium

Arrows indicate the presumed *direction of neuron migration* from neuroepithelial sources.

PLATE 35A
CR 31 mm, GW 9.5, C9226
Frontal/Horizontal, Section 170

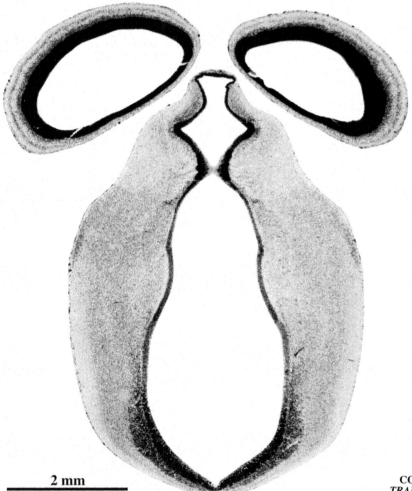

2 mm

LAYERS OF THE
CORTICAL *STRATIFIED*
TRANSITIONAL FIELD (STF)

STF1 Superficial fibrous layer with
an early developmental stage
(t1) when many cells are
migrating through it,
followed by a late stage *(t2)*
with sparse cells. Endures as
the subcortical white matter.

STF5 Deep cellular layer that is
prominent during the first
trimester, the first sojourn
zone to appear outside the
germinal matrix.

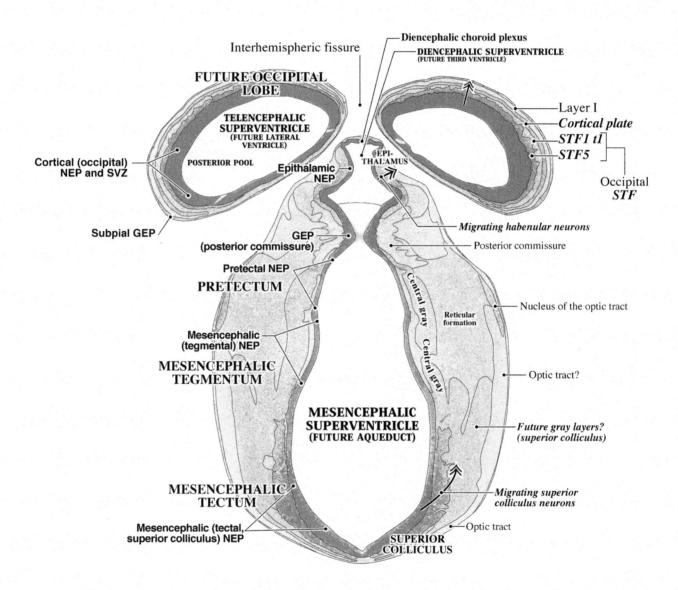

Interhemispheric fissure

Diencephalic choroid plexus

DIENCEPHALIC SUPERVENTRICLE
(FUTURE THIRD VENTRICLE)

FUTURE OCCIPITAL LOBE

Layer I

Cortical plate

TELENCEPHALIC SUPERVENTRICLE
(FUTURE LATERAL VENTRICLE)

STF1 tĪ

STF5

Cortical (occipital) NEP and SVZ

POSTERIOR POOL

Epithalamic NEP

EPI-THALAMUS

Occipital *STF*

Subpial GEP

Migrating habenular neurons

GEP (posterior commissure)

Posterior commissure

Pretectal NEP

PRETECTUM

Central gray

Nucleus of the optic tract

Reticular formation

Mesencephalic (tegmental) NEP

Central gray

MESENCEPHALIC TEGMENTUM

Optic tract?

MESENCEPHALIC SUPERVENTRICLE
(FUTURE AQUEDUCT)

Future gray layers?
(superior colliculus)

MESENCEPHALIC TECTUM

Migrating superior colliculus neurons

Optic tract

Mesencephalic (tectal, superior colliculus) NEP

SUPERIOR COLLICULUS

FONT KEY:
VENTRICULAR DIVISIONS - CAPITALS
Germinal zone - Helvetica bold
Transient structure - Times bold italic
Permanent structure - Times Roman or **Bold**

ABBREVIATIONS:
GEP - Glioepithelium
NEP - Neuroepithelium

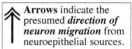

Arrows indicate the presumed *direction of neuron migration* from neuroepithelial sources.

PLATE 36A

CR 31 mm, GW 9.5, C9226
Frontal/Horizontal, Section 158

2 mm

LAYERS OF THE
CORTICAL *STRATIFIED*
TRANSITIONAL FIELD (STF)

STF1 Superficial fibrous layer with
an early developmental stage
(t1) when many cells are
migrating through it,
followed by a late stage *(t2)*
with sparse cells. Endures as
the subcortical white matter.

STF5 Deep cellular layer that is
prominent during the first
trimester, the first sojourn
zone to appear outside the
germinal matrix.

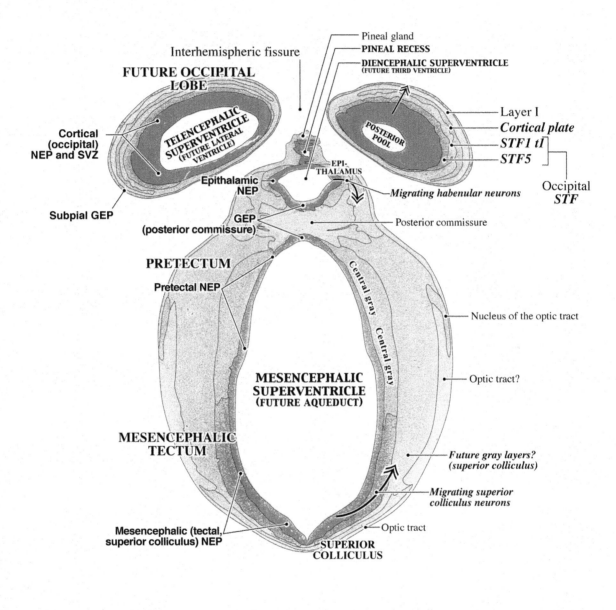

Interhemispheric fissure

Pineal gland
PINEAL RECESS
DIENCEPHALIC SUPERVENTRICLE
(FUTURE THIRD VENTRICLE)

FUTURE OCCIPITAL LOBE

Layer I
Cortical plate
STF1 tI
STF5

Cortical (occipital) NEP and SVZ

TELENCEPHALIC SUPERVENTRICLE (FUTURE LATERAL VENTRICLE)

POSTERIOR POOL

Occipital *STF*

Epithalamic NEP

EPI-THALAMUS

Migrating habenular neurons

Subpial GEP

GEP (posterior commissure)

Posterior commissure

PRETECTUM

Pretectal NEP

Nucleus of the optic tract

Central gray

Central gray

MESENCEPHALIC SUPERVENTRICLE (FUTURE AQUEDUCT)

Optic tract?

MESENCEPHALIC TECTUM

Future gray layers? (superior colliculus)

Migrating superior colliculus neurons

Mesencephalic (tectal, superior colliculus) NEP

Optic tract

SUPERIOR COLLICULUS

FONT KEY:
VENTRICULAR DIVISIONS – CAPITALS
Germinal zone - Helvetica bold
Transient structure - Times bold italic
Permanent structure - Times Roman or **Bold**

ABBREVIATIONS:
GEP - Glioepithelium
NEP - Neuroepithelium
SVZ - Subventricular zone

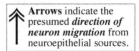

Arrows indicate the presumed *direction of neuron migration* from neuroepithelial sources.

PLATE 37A

CR 31 mm, GW 9.5, C9226, Frontal/Horizontal Section 564
CEREBRAL CORTEX

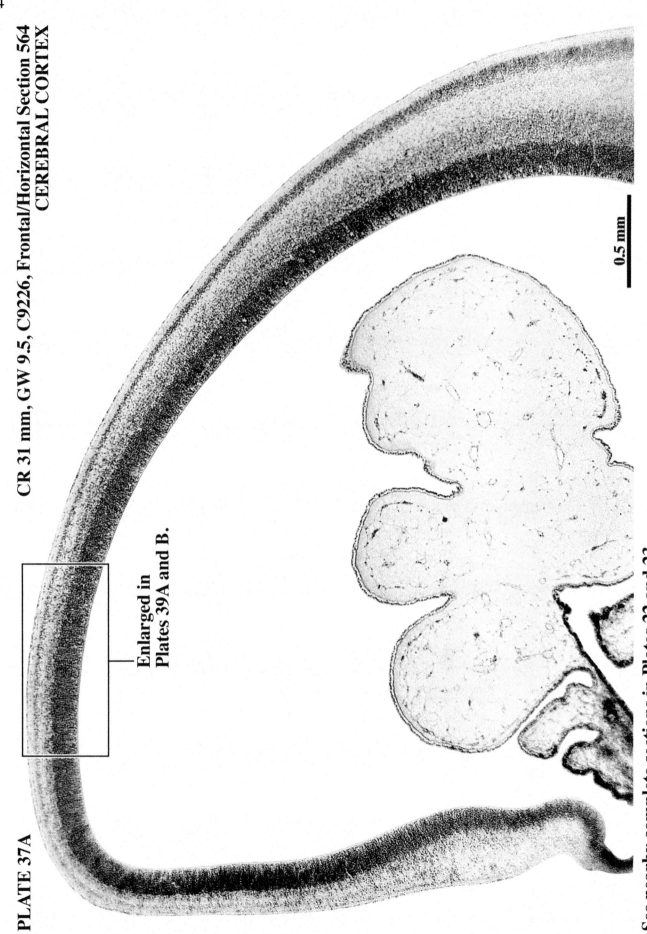

Enlarged in
Plates 39A and B.

0.5 mm

See nearby complete sections in Plates 22 and 23.

PLATE 37B

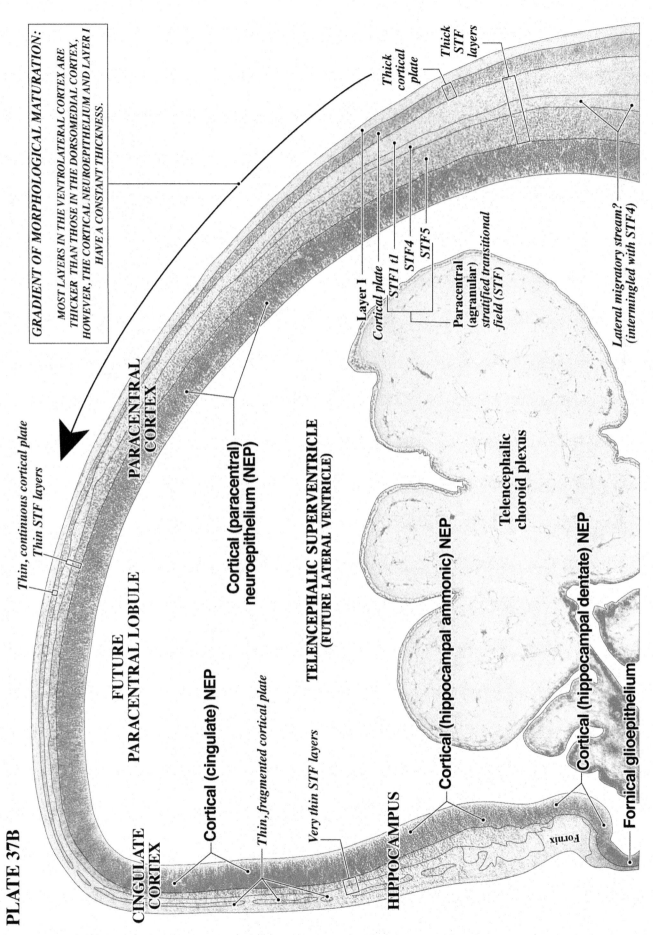

GRADIENT OF MORPHOLOGICAL MATURATION:

MOST LAYERS IN THE VENTROLATERAL CORTEX ARE THICKER THAN THOSE IN THE DORSOMEDIAL CORTEX, HOWEVER, THE CORTICAL NEUROEPITHELIUM AND LAYER 1 HAVE A CONSTANT THICKNESS.

Thick cortical plate

Thick STF layers

Layer I

Cortical plate

STF1 t1

STF4

STF5

Paracentral (agranular) stratified transitional field (STF)

Lateral migratory stream? (intermingled with STF4)

PARACENTRAL CORTEX

Cortical (paracentral) neuroepithelium (NEP)

TELENCEPHALIC SUPERVENTRICLE (FUTURE LATERAL VENTRICLE)

Telencephalic choroid plexus

FUTURE PARACENTRAL LOBULE

Cortical (cingulate) NEP

Thin, fragmented cortical plate

Very thin STF layers

Cortical (hippocampal ammonic) NEP

Cortical (hippocampal dentate) NEP

Thin, continuous cortical plate
Thin STF layers

CINGULATE CORTEX

HIPPOCAMPUS

Fornix

Fornical glioepithelium

PLATE 38A

CR 31 mm, GW 9.5, C9226, Frontal/ Horizontal
PARACENTRAL CEREBRAL CORTEX

Section
564

Section
570

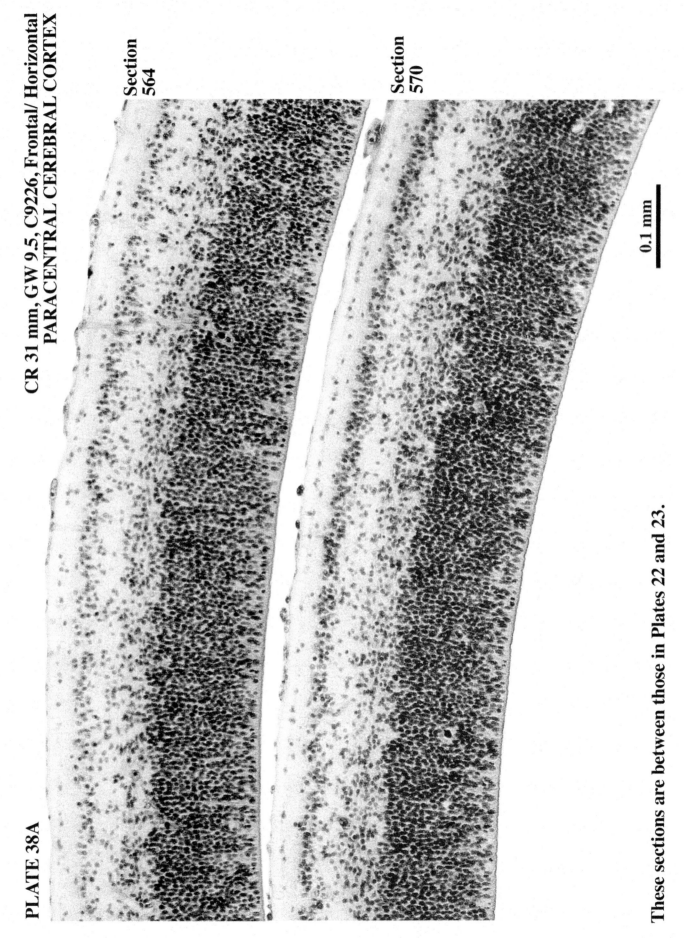

0.1 mm

These sections are between those in Plates 22 and 23.

97

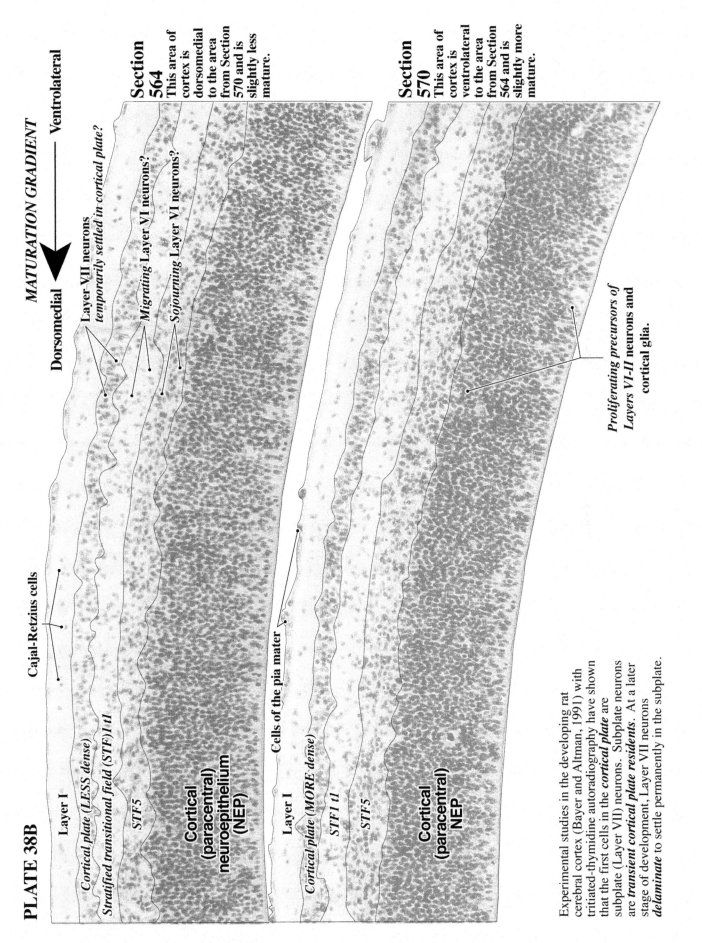

PLATE 38B

Cajal-Retzius cells

MATURATION GRADIENT

Dorsomedial → Ventrolateral

Section 564
This area of cortex is dorsomedial to the area from Section 570 and is slightly less mature.

Layer-VII neurons temporarily settled in cortical plate?

Migrating Layer VI neurons?

Sojourning Layer VI neurons?

Layer I
Cortical plate (LESS dense)
Stratified transitional field (STF)1 t1
STF5

Cortical (paracentral) neuroepithelium (NEP)

Section 570
This area of cortex is ventrolateral to the area from Section 564 and is slightly more mature.

Cells of the pia mater

Layer I
Cortical plate (MORE dense)
STF1 t1
STF5

Cortical (paracentral) NEP

Proliferating precursors of Layers VI-II neurons and cortical glia.

Experimental studies in the developing rat cerebral cortex (Bayer and Altman, 1991) with tritiated-thymidine autoradiography have shown that the first cells in the **cortical plate** are subplate (Layer VII) neurons. Subplate neurons are *transient cortical plate residents*. At a later stage of development, Layer VII neurons *delaminate* to settle permanently in the subplate.

PART IV: C145
CR 33 mm (GW 9.6)
Sagittal

Specimen C145 from the Carnegie Collection is a male with a crown-rump length (CR) of 33 mm estimated to be at gestational week (GW) 9.6. The entire fetus was cut in the sagittal plane. Information on the date of specimen collection, fixative, section thickness, and embedding medium was not available to us. The sections are thick (probably between 50 and 100 μm) and appear to be embedded in celloidin. Since there is no photograph of C145's brain before histological processing, a specimen from Hochstetter (1919) that is comparable in age to C145 is used to show external brain features at GW8 (**A, Figure 16**). C145's brain structures are easier to understand because sections are closely parallel to the midline; **Figure 16** shows the slight rotations in horizontal (**B**) and vertical (**C**) dimensions. Photographs of 10 sections are illustrated at low magnification in four parts (**Plates 39A-D** to **Plates 48A-D**). The A/B parts show the brain in place in the skull; the C/D parts show only the brain (and some peripheral ganglia) at slightly-higher magnification. **Plates 49A-B** to **64A-B** show high-magnification views of various parts of the brain. All of the high-magnification plates are rotated 90 degrees (landscape orientation) to show photographs at higher magnification in the available page space.

The sagittal section plane in the midline shows off the large volume of the brain ventricles when compared to the brain parenchyma (areas where neurons migrate, settle, and differentiate). The largest structure in each of the brain's major subdivisions is the superventricles in their cores. For example, the telencephalon is largely occupied by the telencephalic superventricle. The thickness of the parenchyma is a key to the degree of maturation of the various brain structures. It is thickest in the medulla, pons, and midbrain tegmentum where most of the neurons have been generated and thinner in the cerebellum, midbrain tectum, and diencephalon. Within the telencephalon, the cerebral cortex has a very thin parenchyma, while the basal telencephalon and parts of the basal ganglia have thick parenchymal components.

Throughout the cerebral cortex, the *neuroepithelium* and *subventricular zone* are prominent and blend together as one dense layer. The *stratified transitional field (STF)* contains *STF1* and *STF5* only in lateral areas. The pronounced anterolateral (more mature) to dorsomedial (less mature) gradient is evident in both the cortical plate and the STF layers. The olfactory bulb is just beginning to evaginate in front of the basal telencephalic neuroepithelium. Neurons are just beginning to migrate in the hippocampus. A massive *neuroepithelium/subventricular zone* overlies the striatum (caudate and putamen) where neurons (and glia) are being generated.

The cerebellum has a thick neuroepithelium, in spite of the fact that both deep nuclear neurons and Purkinje cells have already been generated (at least, according to our data in rats). It is suggested that the cerebellar neuroepithelium contains the precursors of Golgi cells that will disperse throughout the granule cell layer in the cerebellar cortex. The deep neurons are superficial layers in the cerebellum, while the Purkinje cells are sojourning in a dense layer at the base of the cerebellar neuroepithelium.. The *external germinal layer (egl)* is a new feature barely visible beneath the pia as one prong of the *germinal trigone* in the dorsal rhombic lip. This dispersed germinal matrix will generate all the microneurons in the cerebellar cortex.

In sections near the midline, the brainstem neuroepithelium varies in thickness. It is thinner in the midbrain tegmentum, pons, and medulla in accordance with an earlier maturation gradient. Most neurons have been generated in these structures and are settling. In the cerebellum (see above), midbrain tectum, and diencephalon, the neuroepithelium is thicker indicating that substantial neurogenesis is happening in these structures, but many neurons have already been produced. However, more lateral sections of the medulla have a thick *precerebellar neuroepithelium*, which is generating pontine gray neurons; some of the pioneer neurons in this population are migrating in the anterior extramural migratory stream.

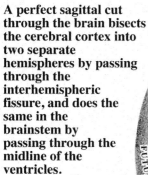

GW9.6 SAGITTAL

A perfect sagittal cut through the brain bisects the cerebral cortex into two separate hemispheres by passing through the interhemispheric fissure, and does the same in the brainstem by passing through the midline of the ventricles.

Sections of C145's brain are very close to the midline both horizontally (+3.48°, top view) and vertically (+3.82°, back view). In each section illustrated on the following pages, the anterior edge of the cortex (top right) is tilted toward the observer, while the medulla and upper spinal cord (bottom) are tilted away from the observer.

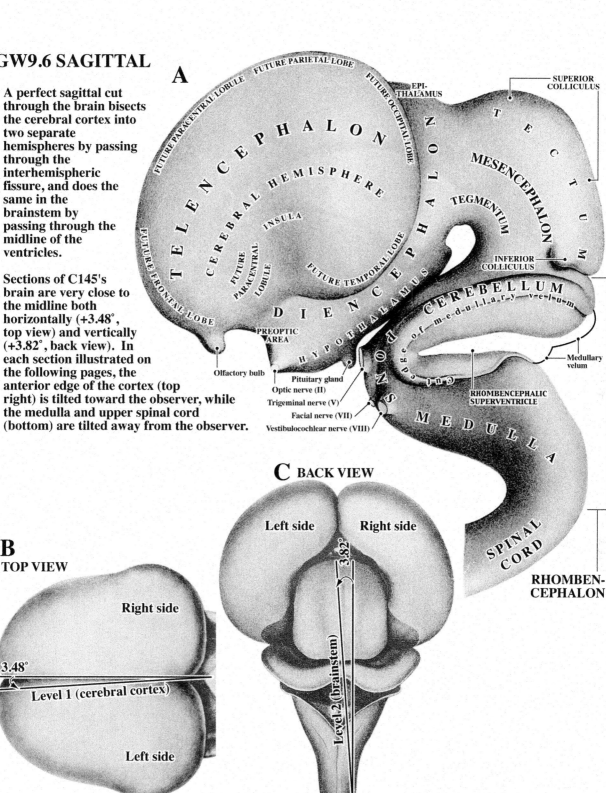

A

FUTURE PARACENTRAL LOBULE
FUTURE PARIETAL LOBE
FUTURE OCCIPITAL LOBE
EPI-THALAMUS
SUPERIOR COLLICULUS

TELENCEPHALON
CEREBRAL HEMISPHERE
INSULA
FUTURE PARACENTRAL LOBULE
FUTURE FRONTAL LOBE
FUTURE TEMPORAL LOBE

DIENCEPHALON
HYPOTHALAMUS
PREOPTIC AREA

MESENCEPHALON
TEGMENTUM
TECTUM
INFERIOR COLLICULUS
CEREBELLUM
of medullary velum
Medullary velum
cut edge of medullary velum

PONS
RHOMBENCEPHALIC SUPERVENTRICLE
MEDULLA

Olfactory bulb
Pituitary gland
Optic nerve (II)
Trigeminal nerve (V)
Facial nerve (VII)
Vestibulocochlear nerve (VIII)

SPINAL CORD
RHOMBEN-CEPHALON

C BACK VIEW

Left side   Right side

3.82°
Level-2 (brainstem)

B
TOP VIEW

Right side

3.48°
Level 1 (cerebral cortex)

Left side

Figure 16. A, the lateral view of the brain and upper cervical spinal cord from a specimen with a crown-rump length of 27 mm (modified from Figure 39, Table VII, Hochstetter, 1919) identifies external features of a brain similar to C145 (CR 33 mm). **B**, top view of the brain with a crown-rump length of 38 mm (modified from Figure 45, Table VIII, Hochstetter, 1919) shows how C145's sections rotate from a line parallel to the horizontal midline in the interhemispheric fissure. **C**, back view of the brain in **B** (modified from Figure 44, Table VIII, Hochstetter, 1919) shows how C145's sections rotate from a line parallel to the vertical midline in the brainstem and upper cervical spinal cord.

CR 33 mm, GW 9.6, C145
Sagittal, Slide 23, Section 2

HEAD STRUCTURES,
MAJOR BRAIN REGIONS,
AND VENTRICULAR
DIVISIONS

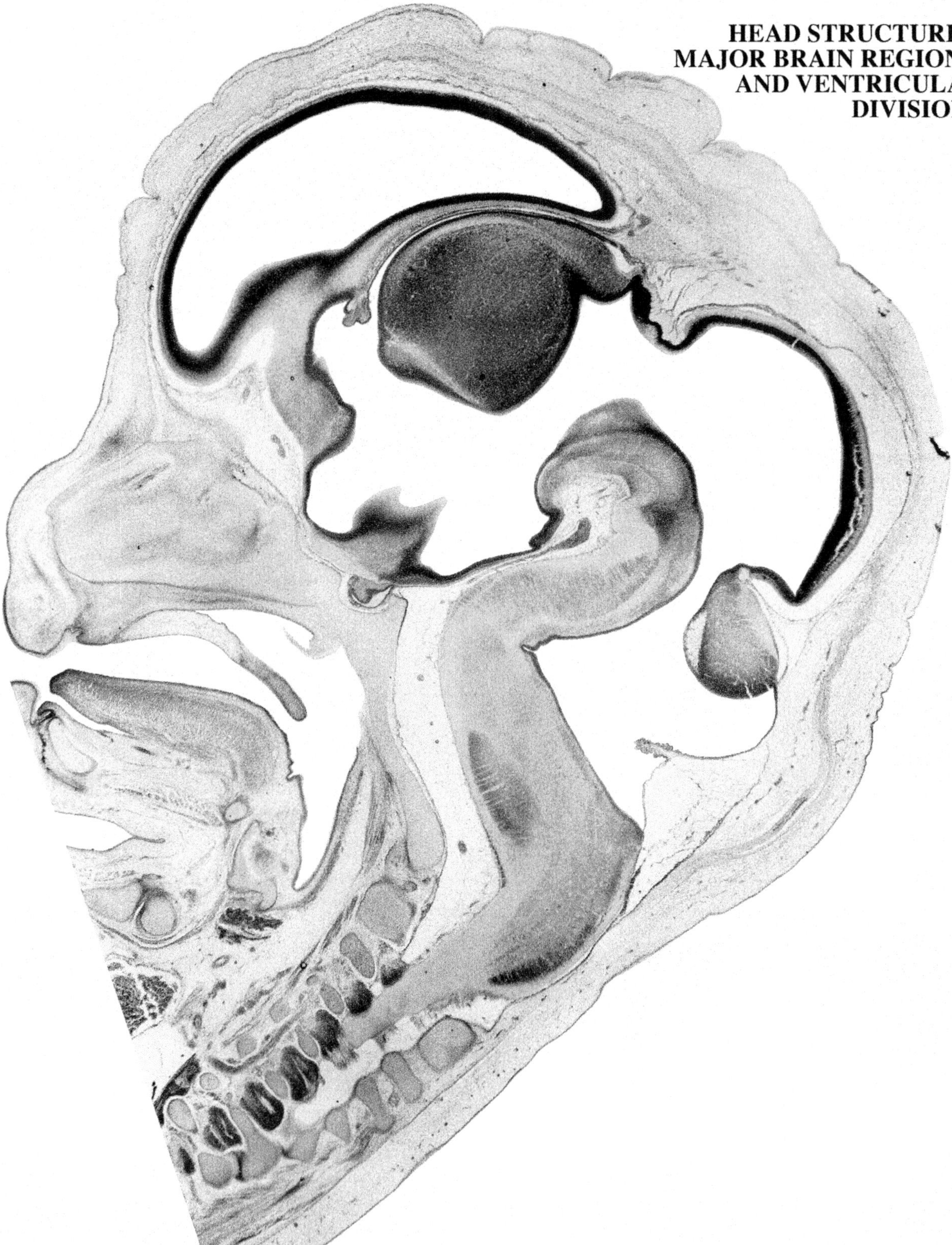

2 mm

Germinal matrix divisions and
differentiating structures are
labeled in Parts C and D of this
plate on the following pages.

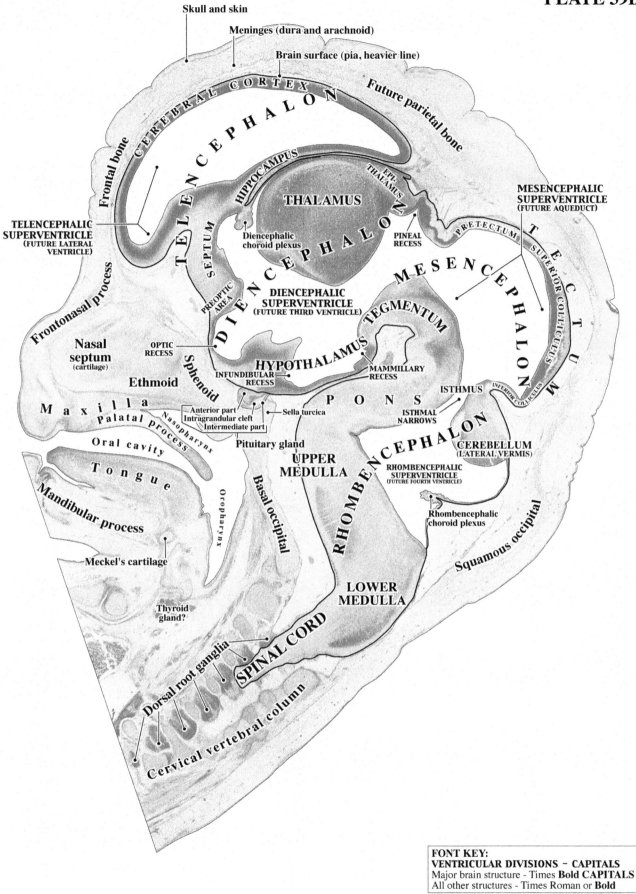

Skull and skin

Meninges (dura and arachnoid)

Brain surface (pia, heavier line)

CEREBRAL CORTEX

Future parietal bone

TELENCEPHALON

Frontal bone

HIPPOCAMPUS

THALAMUS

EPI. THALAMUS

MESENCEPHALIC SUPERVENTRICLE
(FUTURE AQUEDUCT)

TELENCEPHALIC SUPERVENTRICLE
(FUTURE LATERAL VENTRICLE)

DIENCEPHALON

Diencephalic choroid plexus

PINEAL RECESS

PRE-TECTUM

T E C T U M

SUPERIOR COLLICULUS

SEPTUM

MESENCEPHALON

Frontonasal process

PREOPTIC AREA

DIENCEPHALIC SUPERVENTRICLE
(FUTURE THIRD VENTRICLE)

TEGMENTUM

Nasal septum
(cartilage)

OPTIC RECESS

HYPOTHALAMUS

MAMMILLARY RECESS

ISTHMUS

INFERIOR COLLICULUS

Sphenoid

INFUNDIBULAR RECESS

Ethmoid

P O N S

ISTHMAL NARROWS

M a x i l l a

Anterior part

Sella turcica

CEREBELLUM
(LATERAL VERMIS)

Palatal process

Nasopharynx

Intragrandular cleft

Intermediate part

Oral cavity

Pituitary gland

UPPER MEDULLA

RHOMBENCEPHALON

RHOMBENCEPHALIC SUPERVENTRICLE
(FUTURE FOURTH VENTRICLE)

T o n g u e

Oropharynx

Mandibular process

Basal occipital

Rhombencephalic choroid plexus

Meckel's cartilage

Squamous occipital

LOWER MEDULLA

Thyroid gland?

SPINAL CORD

Dorsal root ganglia

Cervical vertebral column

FONT KEY:
VENTRICULAR DIVISIONS – CAPITALS
Major brain structure - Times **Bold CAPITALS**
All other structures - Times Roman or **Bold**

PLATE 39C

CR 33 mm, GW 9.6, C145
Sagittal, Slide 23, Section 2

**GERMINAL MATRIX DIVISIONS
AND DIFFERENTIATING
BRAIN STRUCTURES**

Left side

Midline ——————————

Right side

BRAINSTEM FLEXURES

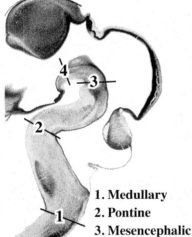

1. Medullary
2. Pontine
3. Mesencephalic
4. Diencephalic

See high-magnification
views of the midbrain
medial to this section in
Plates 55 to 56A and B,
of the pons and medulla
from this section in
Plates 59A and B.

The head, major brain
structures, and ventricular
divisions are labeled in Parts
A and B of this plate on the
preceding pages.

2 mm

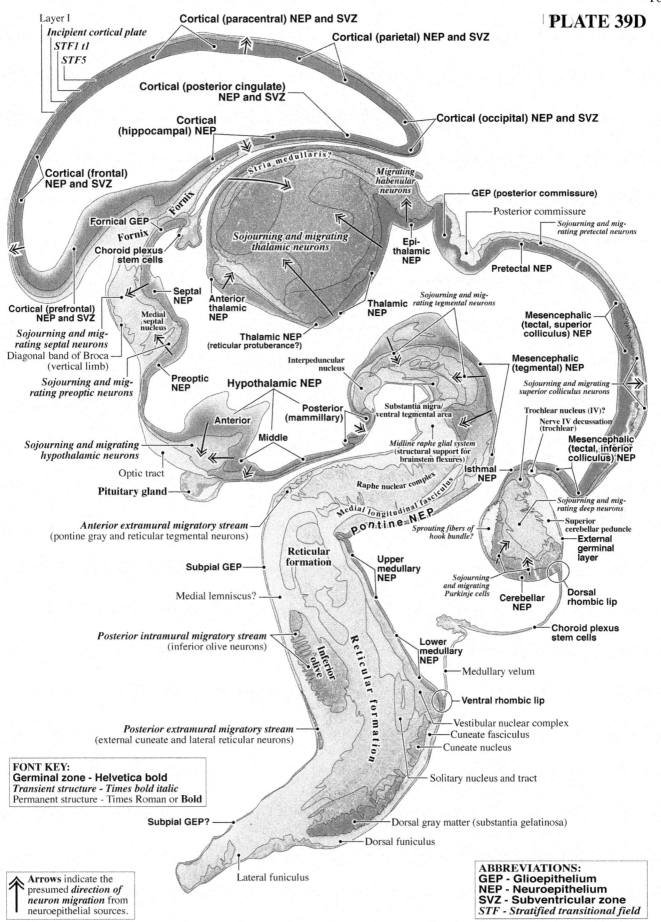

PLATE 39D

Layer I
Incipient cortical plate
STF1 t1
STF5

Cortical (paracentral) NEP and SVZ

Cortical (parietal) NEP and SVZ

Cortical (posterior cingulate) NEP and SVZ

Cortical (hippocampal) NEP

Cortical (occipital) NEP and SVZ

Cortical (frontal) NEP and SVZ

Stria medullaris?

Migrating habenular neurons

GEP (posterior commissure)

Posterior commissure

Fornix

Fornical GEP

Fornix

Choroid plexus stem cells

Sojourning and migrating thalamic neurons

Epi-thalamic NEP

Sojourning and migrating pretectal neurons

Pretectal NEP

Cortical (prefrontal) NEP and SVZ

Septal NEP

Anterior thalamic NEP

Thalamic NEP

Sojourning and migrating tegmental neurons

Mesencephalic (tectal, superior colliculus) NEP

Sojourning and migrating septal neurons

Medial septal nucleus

Thalamic NEP (reticular protuberance?)

Mesencephalic (tegmental) NEP

Diagonal band of Broca (vertical limb)

Interpeduncular nucleus

Sojourning and migrating superior colliculus neurons

Sojourning and migrating preoptic neurons

Preoptic NEP

Hypothalamic NEP

Substantia nigra/ventral tegmental area

Trochlear nucleus (IV)?

Nerve IV decussation (trochlear)

Posterior (mammillary)

Anterior

Middle

Midline raphe glial system (structural support for brainstem flexures)

Mesencephalic (tectal, inferior colliculus) NEP

Sojourning and migrating hypothalamic neurons

Isthmal NEP

Sojourning and migrating deep neurons

Optic tract

Raphe nuclear complex

Pituitary gland

Medial longitudinal fasciculus

Sprouting fibers of hook bundle?

Superior cerebellar peduncle

External germinal layer

Anterior extramural migratory stream (pontine gray and reticular tegmental neurons)

Pontine NEP

Reticular formation

Upper medullary NEP

Sojourning and migrating Purkinje cells

Cerebellar NEP

Dorsal rhombic lip

Subpial GEP

Medial lemniscus?

Choroid plexus stem cells

Posterior intramural migratory stream (inferior olive neurons)

Inferior olive

Reticular formation

Lower medullary NEP

Medullary velum

Ventral rhombic lip

Vestibular nuclear complex

Cuneate fasciculus

Cuneate nucleus

Posterior extramural migratory stream (external cuneate and lateral reticular neurons)

Solitary nucleus and tract

FONT KEY:
Germinal zone - Helvetica bold
Transient structure - Times bold italic
Permanent structure - Times Roman or **Bold**

Subpial GEP?

Dorsal gray matter (substantia gelatinosa)

Dorsal funiculus

Lateral funiculus

Arrows indicate the presumed *direction of neuron migration* from neuroepithelial sources.

ABBREVIATIONS:
GEP - Glioepithelium
NEP - Neuroepithelium
SVZ - Subventricular zone
STF - Stratified transitional field

CR 33 mm, GW 9.6, C145
Sagittal, Slide 22, Section 2

HEAD STRUCTURES,
MAJOR BRAIN REGIONS,
AND VENTRICULAR
DIVISIONS

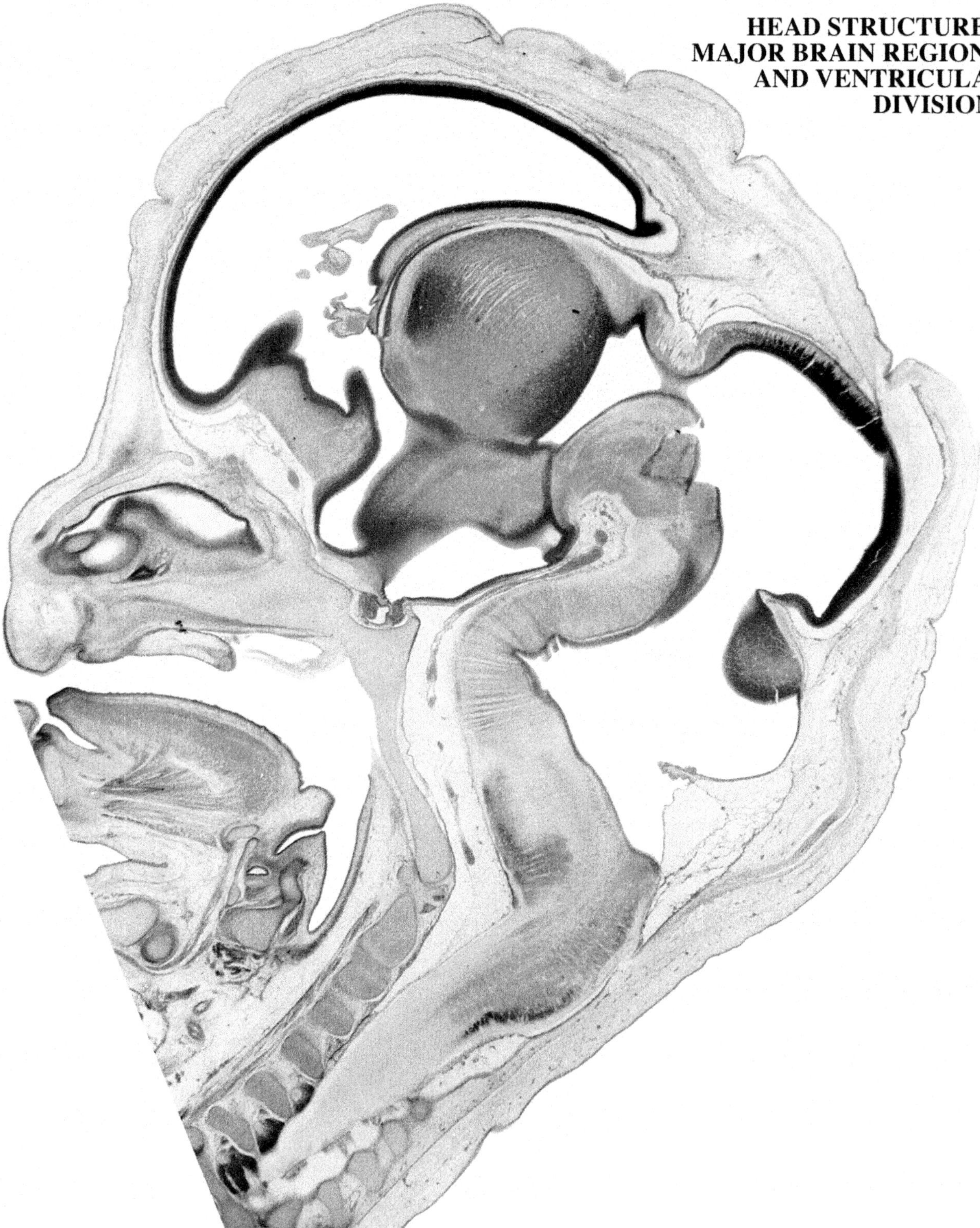

2 mm

Germinal matrix divisions, and differentiating
structures are labeled in Parts C and D of this
plate on the following pages.

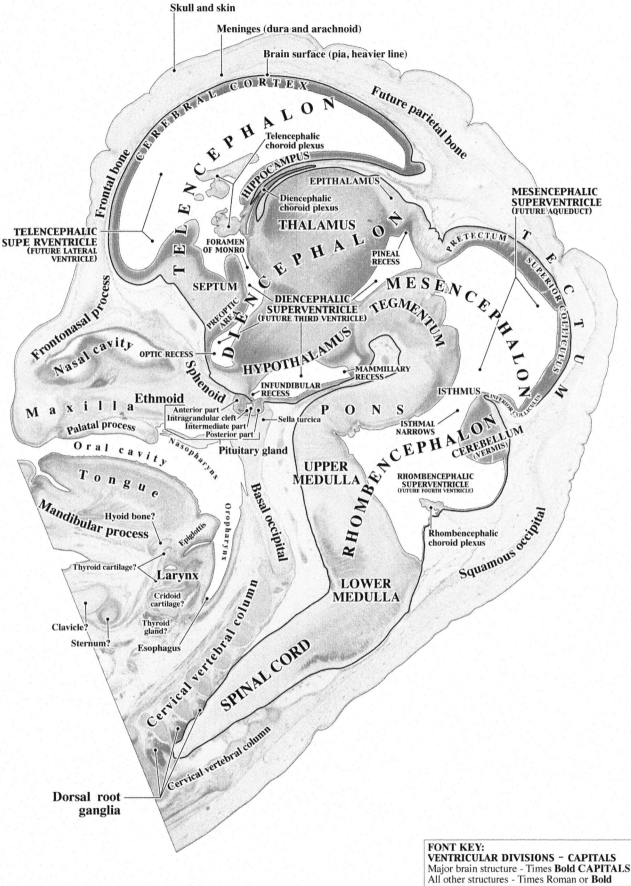

Skull and skin

Meninges (dura and arachnoid)

Brain surface (pia, heavier line)

C E R E B R A L   C O R T E X

Future parietal bone

T E L E N C E P H A L O N

Frontal bone

Telencephalic
choroid plexus

HIPPOCAMPUS

EPITHALAMUS

MESENCEPHALIC
SUPERVENTRICLE
(FUTURE AQUEDUCT)

Diencephalic
choroid plexus

THALAMUS

D I E N C E P H A L O N

PRETECTUM

T E C T U M

SUPERIOR COLLICULUS

TELENCEPHALIC
SUPE RVENTRICLE
(FUTURE LATERAL
VENTRICLE)

FORAMEN
OF MONRO

PINEAL
RECESS

M E S E N C E P H A L O N

SEPTUM

Frontonasal process

PREOPTIC
AREA

DIENCEPHALIC
SUPERVENTRICLE
(FUTURE THIRD VENTRICLE)

TEGMENTUM

Nasal cavity

OPTIC RECESS

HYPOTHALAMUS

MAMMILLARY
RECESS

Sphenoid

ISTHMUS

INFERIOR COLLICULUS

M a x i l l a   Ethmoid

INFUNDIBULAR
RECESS

P O N S

ISTHMAL
NARROWS

Anterior part
Intragrandular cleft
Intermediate part
Posterior part

Sella turcica

CEREBELLUM
(VERMIS)

Palatal process

Nasopharynx

Pituitary gland

O r a l   c a v i t y

T o n g u e

UPPER
MEDULLA

RHOMBENCEPHALIC
SUPERVENTRICLE
(FUTURE FOURTH VENTRICLE)

Mandibular process

Hyoid bone?

Oropharynx

Epiglottis

Basal occipital

R H O M B E N C E P H A L O N

Thyroid cartilage?

Larynx

Rhombencephalic
choroid plexus

Cridoid
cartilage?

Squamous occipital

Clavicle?

Thyroid
gland?

LOWER
MEDULLA

Sternum?

Esophagus

Cervical vertebral column

SPINAL CORD

Cervical vertebral column

Dorsal root
ganglia

FONT KEY:
VENTRICULAR DIVISIONS – CAPITALS
Major brain structure - Times **Bold CAPITALS**
All other structures - Times Roman or **Bold**

PLATE 40C

**GERMINAL MATRIX DIVISIONS
AND DIFFERENTIATING
BRAIN STRUCTURES**

Left side

Midline

Right side

BRAINSTEM FLEXURES

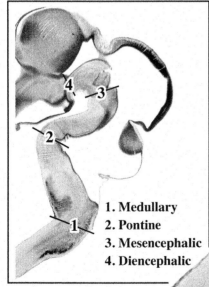

1. Medullary
2. Pontine
3. Mesencephalic
4. Diencephalic

See high-magnification views
of the hypothalamus and
basal telencephalon in Plates
50A and B, of the midbrain,
pons, and medulla in Plates
60A and B.

The head, major brain
structures, and ventricular
divisions are labeled in Parts
A and B of this plate on the
preceding pages.

2 mm

PLATE 40D

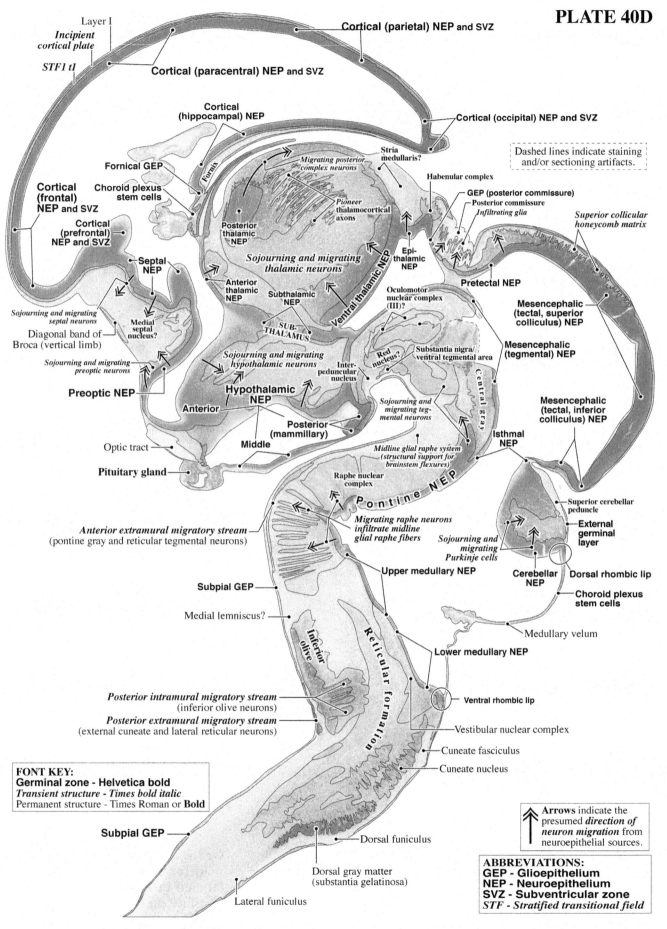

Layer I
Incipient cortical plate
STF1 t1

Cortical (parietal) NEP and SVZ

Cortical (paracentral) NEP and SVZ

Cortical (hippocampal) NEP

Cortical (occipital) NEP and SVZ

Fornical GEP

Fornix

Stria medullaris?

Dashed lines indicate staining and/or sectioning artifacts.

Migrating posterior complex neurons

Habenular complex

GEP (posterior commissure)
Posterior commissure
Infiltrating glia

Superior collicular honeycomb matrix

Cortical (frontal) NEP and SVZ

Choroid plexus stem cells

Posterior thalamic NEP

Pioneer thalamocortical axons

Cortical (prefrontal) NEP and SVZ

Sojourning and migrating thalamic neurons

Epithalamic NEP

Ventral thalamic NEP

Pretectal NEP

Septal NEP

Anterior thalamic NEP

Subthalamic NEP

Oculomotor nuclear complex (III)?

Mesencephalic (tectal, superior colliculus) NEP

Sojourning and migrating septal neurons

Medial septal nucleus?

SUB-THALAMUS

Mesencephalic (tegmental) NEP

Diagonal band of Broca (vertical limb)

Sojourning and migrating hypothalamic neurons

Interpeduncular nucleus

Red nucleus?

Substantia nigra/ ventral tegmental area

Central gray

Sojourning and migrating preoptic neurons

Mesencephalic (tectal, inferior colliculus) NEP

Preoptic NEP

Hypothalamic NEP

Sojourning and migrating tegmental neurons

Anterior

Posterior (mammillary)

Isthmal NEP

Optic tract

Middle

Midline glial raphe system (structural support for brainstem flexures)

Raphe nuclear complex

Pituitary gland

Pontine NEP

Superior cerebellar peduncle

External germinal layer

Anterior extramural migratory stream (pontine gray and reticular tegmental neurons)

Migrating raphe neurons infiltrate midline glial raphe fibers

Sojourning and migrating Purkinje cells

Cerebellar NEP

Dorsal rhombic lip

Upper medullary NEP

Choroid plexus stem cells

Subpial GEP

Medullary velum

Medial lemniscus?

Inferior olive

Reticular formation

Lower medullary NEP

Posterior intramural migratory stream (inferior olive neurons)

Ventral rhombic lip

Posterior extramural migratory stream (external cuneate and lateral reticular neurons)

Vestibular nuclear complex

Cuneate fasciculus

Cuneate nucleus

FONT KEY:
Germinal zone - Helvetica bold
Transient structure - Times bold italic
Permanent structure - Times Roman or **Bold**

Subpial GEP

Dorsal funiculus

Dorsal gray matter (substantia gelatinosa)

Lateral funiculus

Arrows indicate the presumed *direction of neuron migration* from neuroepithelial sources.

ABBREVIATIONS:
GEP - Glioepithelium
NEP - Neuroepithelium
SVZ - Subventricular zone
STF - Stratified transitional field

PLATE 41A

**HEAD STRUCTURES,
MAJOR BRAIN REGIONS,
AND VENTRICULAR
DIVISIONS**

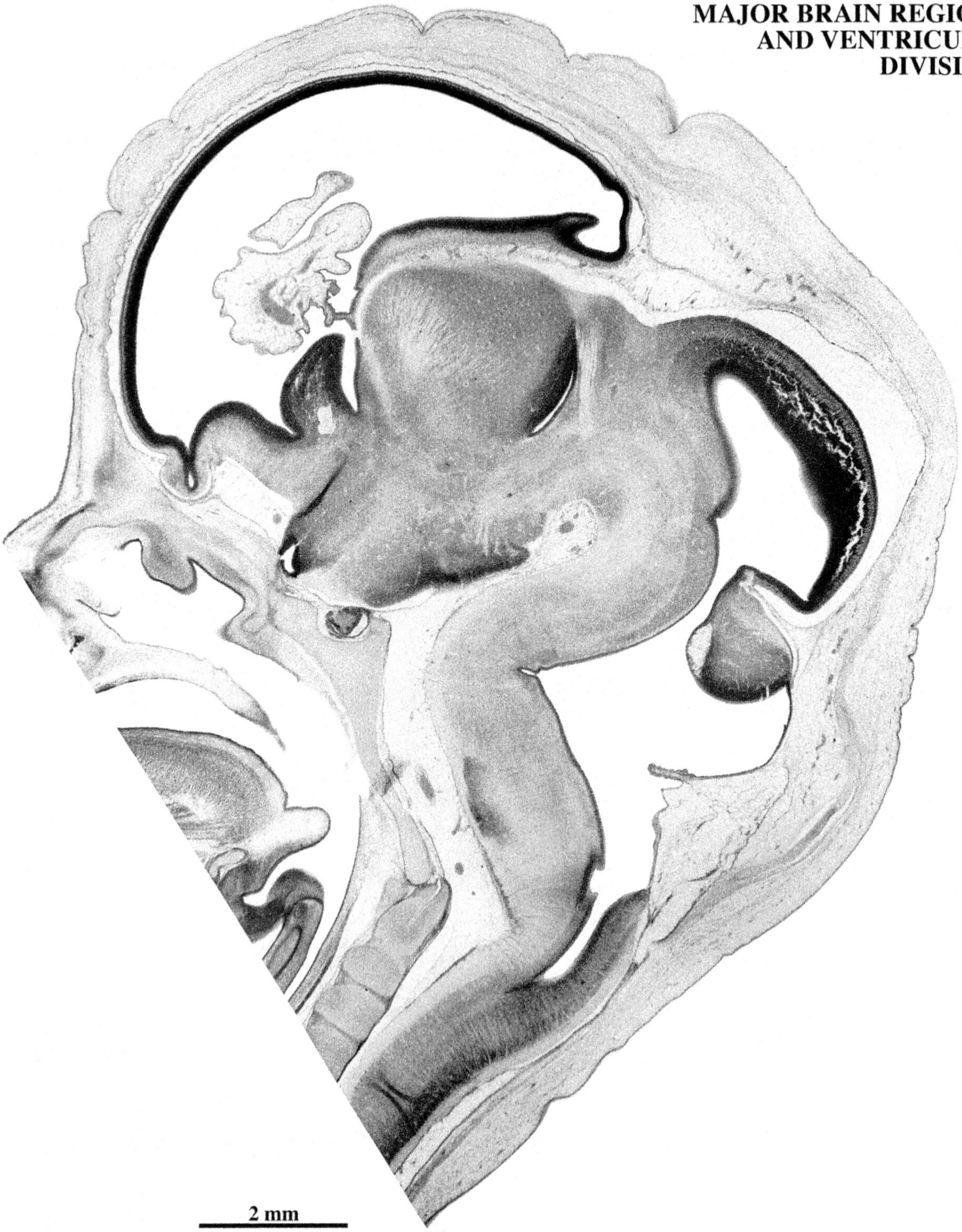

2 mm

**Germinal matrix divisions, and
differentiating structures are
labeled in Parts C and D of this
plate on the following pages.**

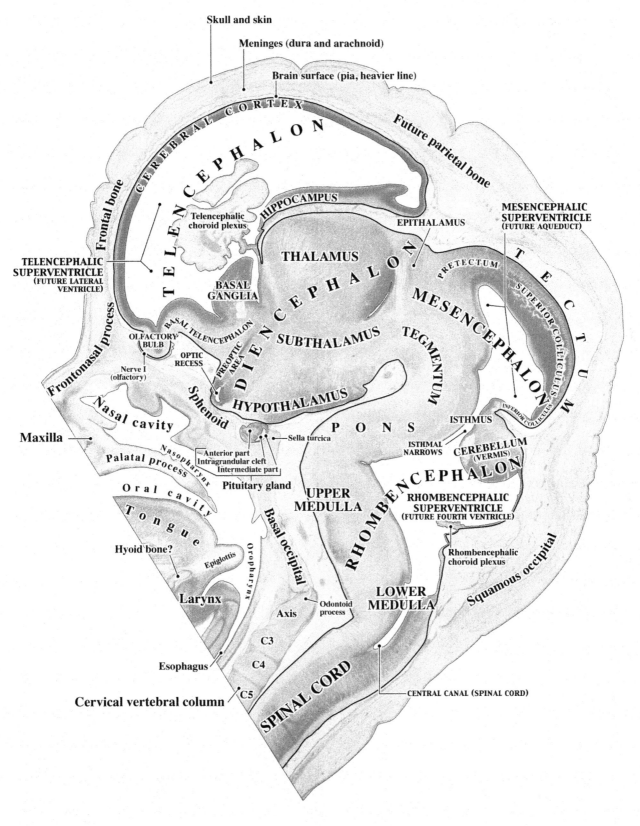

Skull and skin

Meninges (dura and arachnoid)

Brain surface (pia, heavier line)

CEREBRAL CORTEX

TELENCEPHALON

Future parietal bone

Frontal bone

HIPPOCAMPUS

Telencephalic choroid plexus

EPITHALAMUS

MESENCEPHALIC SUPERVENTRICLE (FUTURE AQUEDUCT)

THALAMUS

PRETECTUM

TECTUM

TELENCEPHALIC SUPERVENTRICLE (FUTURE LATERAL VENTRICLE)

DIENCEPHALON

MESENCEPHALON

SUPERIOR COLLICULUS

BASAL GANGLIA

BASAL TELENCEPHALON

OLFACTORY BULB

SUBTHALAMUS

TEGMENTUM

OPTIC RECESS

PREOPTIC AREA

Frontonasal process

Nerve I (olfactory)

Nasal cavity

Sphenoid

HYPOTHALAMUS

INFERIOR COLLICULUS

ISTHMUS

Maxilla

Sella turcica

PONS

ISTHMAL NARROWS

CEREBELLUM (VERMIS)

Nasopharynx

Palatal process

Anterior part
Intragrandular cleft
Intermediate part

Pituitary gland

Oral cavity

Tongue

Hyoid bone?

Epiglottis

UPPER MEDULLA

RHOMBENCEPHALIC SUPERVENTRICLE (FUTURE FOURTH VENTRICLE)

Larynx

Oropharynx

Basal occipital

Rhombencephalic choroid plexus

RHOMBENCEPHALON

Squamous occipital

Axis

Odontoid process

LOWER MEDULLA

C3

C4

Esophagus

C5

Cervical vertebral column

SPINAL CORD

CENTRAL CANAL (SPINAL CORD)

FONT KEY:
VENTRICULAR DIVISIONS – CAPITALS
Major brain structure - Times **Bold CAPITALS**
All other structures - Times Roman or **Bold**

CR 33 mm, GW 9.6, C145
Sagittal, Slide 20, Section 2

GERMINAL MATRIX DIVISIONS
AND DIFFERENTIATING
BRAIN STRUCTURES

The head, major brain
structures, and ventricular
divisions are labeled in Parts
A and B of this plate on the
preceding pages.

BRAINSTEM FLEXURES

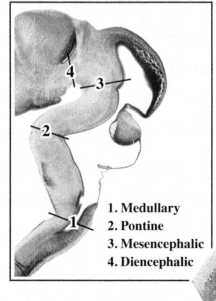

1. Medullary
2. Pontine
3. Mesencephalic
4. Diencephalic

Left side — Midline
Right side

See high-magnification
views of the diencephalon
and basal telencephalon in
Plates 51A and B, the
midbrain, pons, and
medulla in Plates 62A and
B, the cerebellum in Plates
57A and B.

2 mm

PLATE 41D

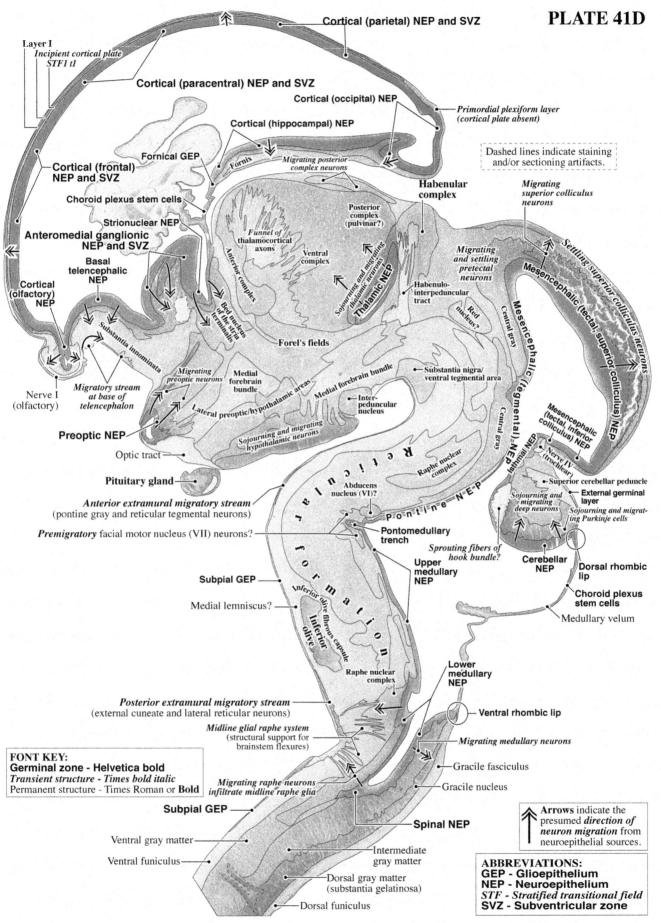

Cortical (parietal) NEP and SVZ

Layer I
Incipient cortical plate
STF1 t1

Cortical (paracentral) NEP and SVZ

Cortical (occipital) NEP

Cortical (hippocampal) NEP

Primordial plexiform layer
(cortical plate absent)

Dashed lines indicate staining
and/or sectioning artifacts.

Fornical GEP

Fornix

Migrating posterior-
complex neurons

Habenular
complex

Migrating
superior colliculus
neurons

Cortical (frontal)
NEP and SVZ

Choroid plexus stem cells

Posterior
complex
(pulvinar?)

Strionuclear NEP

Funnel of
thalamocortical
axons

Ventral
complex

Migrating
and settling
pretectal
neurons

Mesencephalic (tectal, superior colliculus) neurons

Settling superior colliculus neurons

Anteromedial ganglionic
NEP and SVZ

Anterior complex

Basal
telencephalic
NEP

Cortical
(olfactory)
NEP

Bed nucleus
of the stria
terminalis

Thalamic NEP

Sojourning and migrating
thalamic neurons

Habenulo-
interpeduncular
tract

Red
nucleus?

Mesencephalic (tegmental) NEP

Central gray

Substantia innominata

Migrating
preoptic neurons

Medial
forebrain
bundle

Medial forebrain bundle

Substantia nigra/
ventral tegmental area

Central gray

Mesencephalic
(tectal, inferior
colliculus) NEP

Nerve I
(olfactory)

Migratory stream
at base of
telencephalon

Lateral preoptic/hypothalamic areas

Inter-
peduncular
nucleus

Isthmal NEP

Nerve IV
(trochlear)

Preoptic NEP

Optic tract

Sojourning and migrating
hypothalamic neurons

R
e
t
i
c
u
l
a
r

Raphe nuclear
complex

Superior cerebellar peduncle

External germinal
layer

Pituitary gland

Anterior extramural migratory stream
(pontine gray and reticular tegmental neurons)

Abducens
nucleus (VI)?

Sojourning and
migrating
deep neurons

Sojourning and migrat-
ing Purkinje cells

Premigratory facial motor nucleus (VII) neurons?

P o n t i n e   N E P

Sprouting fibers of
hook bundle?

Cerebellar
NEP

Dorsal rhombic
lip

Pontomedullary
trench

Upper
medullary
NEP

Subpial GEP

f
o
r
m
a
t
i
o
n

Inferior olive fibrous capsule

Choroid plexus
stem cells

Medial lemniscus?

Inferior
olive

Medullary velum

Lower
medullary
NEP

Raphe nuclear
complex

Posterior extramural migratory stream
(external cuneate and lateral reticular neurons)

Ventral rhombic lip

Midline glial raphe system
(structural support for
brainstem flexures)

Migrating medullary neurons

Gracile fasciculus

Subpial GEP

Migrating raphe neurons
infiltrate midline raphe glia

Gracile nucleus

FONT KEY:
Germinal zone - Helvetica bold
Transient structure - Times bold italic
Permanent structure - Times Roman or **Bold**

Spinal NEP

Ventral gray matter

Ventral funiculus

Intermediate
gray matter

Dorsal gray matter
(substantia gelatinosa)

Dorsal funiculus

Arrows indicate the
presumed *direction of*
neuron migration from
neuroepithelial sources.

ABBREVIATIONS:
GEP - Glioepithelium
NEP - Neuroepithelium
STF - Stratified transitional field
SVZ - Subventricular zone

CR 33 mm, GW 9.6, C145
Sagittal, Slide 19, Section 2

HEAD STRUCTURES,
MAJOR BRAIN REGIONS,
AND VENTRICULAR
DIVISIONS

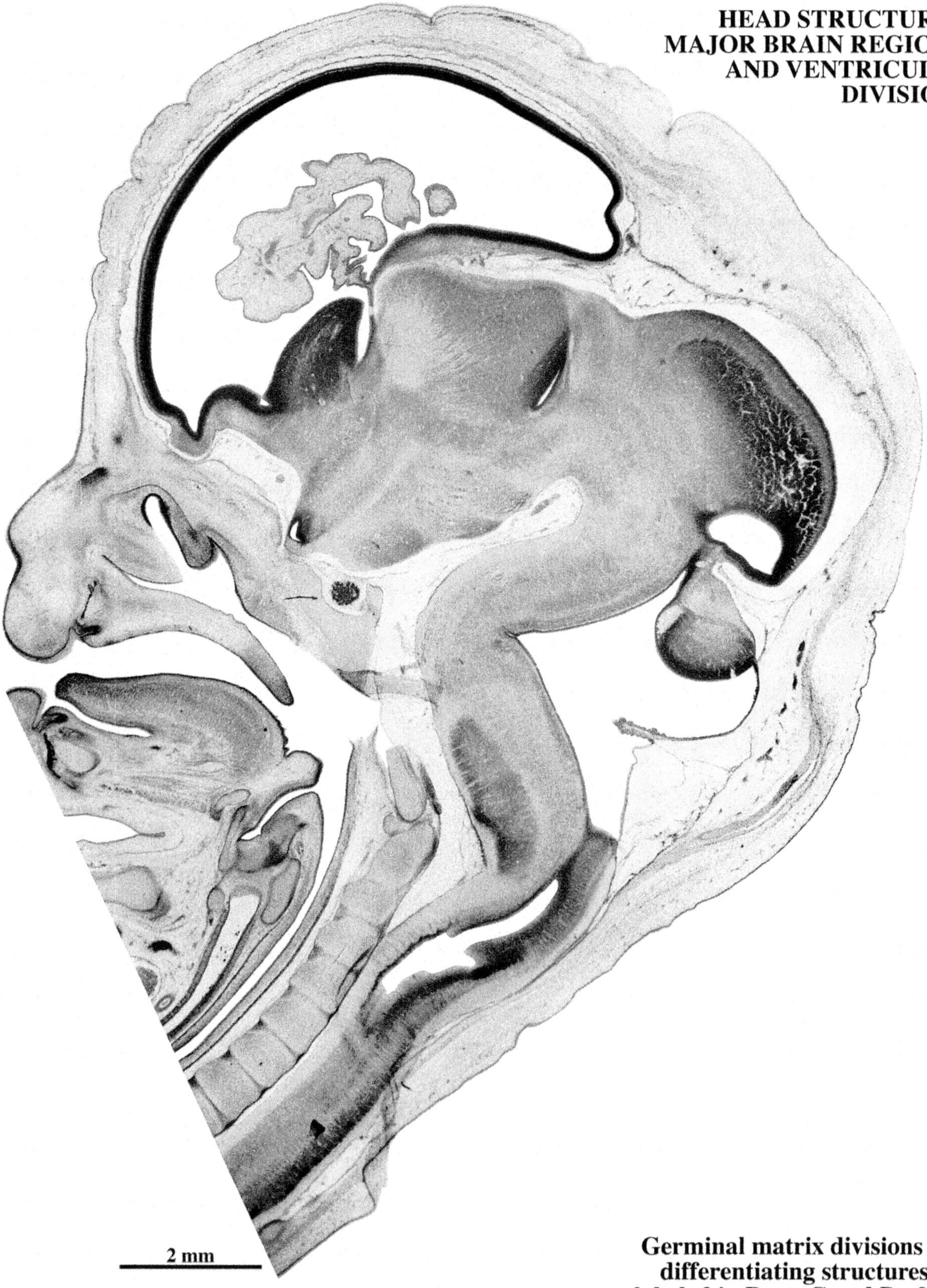

2 mm

Germinal matrix divisions and
differentiating structures are
labeled in Parts C and D of this
plate on the following pages.

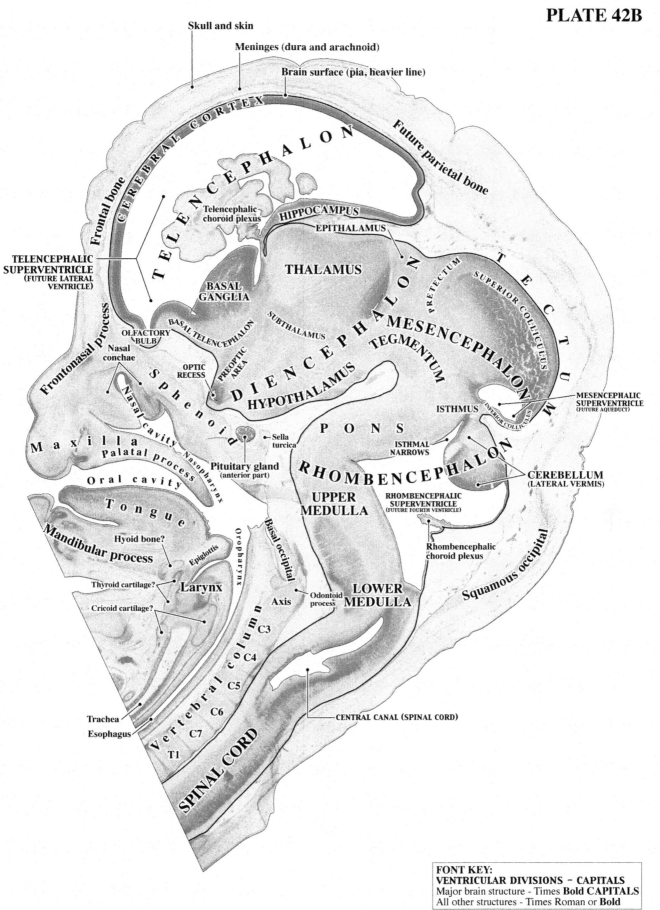

Skull and skin

Meninges (dura and arachnoid)

Brain surface (pia, heavier line)

CEREBRAL CORTEX

TELENCEPHALON

Future parietal bone

Frontal bone

Telencephalic choroid plexus

HIPPOCAMPUS

EPITHALAMUS

THALAMUS

TELENCEPHALIC SUPERVENTRICLE (FUTURE LATERAL VENTRICLE)

BASAL GANGLIA

DIENCEPHALON

TECTUM

PRETECTUM

SUPERIOR COLLICULUS

BASAL TELENCEPHALON

SUBTHALAMUS

MESENCEPHALON

OLFACTORY BULB

Frontonasal process

Nasal conchae

PREOPTIC AREA

TEGMENTUM

OPTIC RECESS

Sphenoid

HYPOTHALAMUS

ISTHMUS

INFERIOR COLLICULUS

MESENCEPHALIC SUPERVENTRICLE (FUTURE AQUEDUCT)

Maxilla

Nasal cavity

Nasopharynx

Sella turcica

PONS

ISTHMAL NARROWS

Palatal process

Pituitary gland (anterior part)

RHOMBENCEPHALON

CEREBELLUM (LATERAL VERMIS)

Oral cavity

UPPER MEDULLA

RHOMBENCEPHALIC SUPERVENTRICLE (FUTURE FOURTH VENTRICLE)

Tongue

Oropharynx

Basal occipital

Hyoid bone?

Rhombencephalic choroid plexus

Mandibular process

Epiglottis

Thyroid cartilage?

Larynx

LOWER MEDULLA

Squamous occipital

Cricoid cartilage?

Axis

Odontoid process

C3

Vertebral column

C4

C5

C6

Trachea

C7

CENTRAL CANAL (SPINAL CORD)

Esophagus

T1

SPINAL CORD

FONT KEY:
VENTRICULAR DIVISIONS – CAPITALS
Major brain structure - Times **Bold CAPITALS**
All other structures - Times Roman or **Bold**

PLATE 42C

**GERMINAL MATRIX DIVI-
SIONS AND DIFFEREN-
TIATING BRAIN
STRUCTURES**

See high-magnification
views of the diencephalon
and basal telencephalon in
Plates 52A and B.

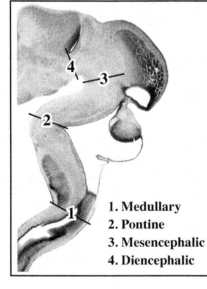

BRAINSTEM FLEXURES

1. Medullary
2. Pontine
3. Mesencephalic
4. Diencephalic

Left side — Midline
Right side

The head, major brain
structures, and ventricular
divisions are labeled in
Parts A and B of this plate
on the preceding pages.

2 mm

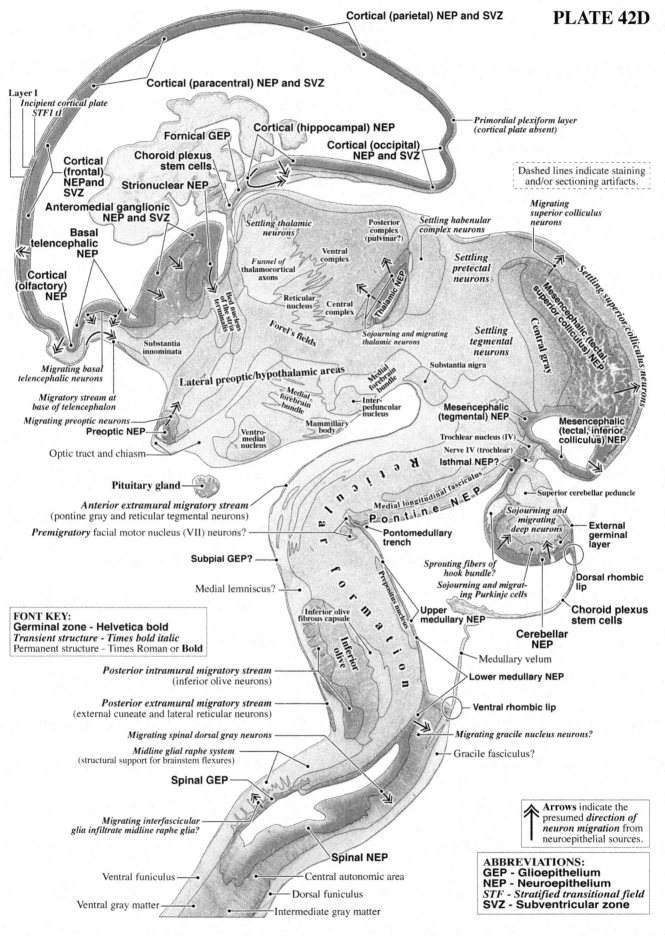

Cortical (parietal) NEP and SVZ

Cortical (paracentral) NEP and SVZ

Layer I
Incipient cortical plate STF1 t1

Fornical GEP

Cortical (hippocampal) NEP

Choroid plexus stem cells

Cortical (occipital) NEP and SVZ

Primordial plexiform layer (cortical plate absent)

Cortical (frontal) NEP and SVZ

Strionuclear NEP

Dashed lines indicate staining and/or sectioning artifacts.

Anteromedial ganglionic NEP and SVZ

Migrating superior colliculus neurons

Basal telencephalic NEP

Settling thalamic neurons

Posterior complex (pulvinar?)

Settling habenular complex neurons

Settling pretectal neurons

Cortical (olfactory) NEP

Ventral complex

Funnel of thalamocortical axons

Thalamic NEP

Mesencephalic (tectal) superior colliculus NEP

Settling superior colliculus neurons

Reticular nucleus

Central complex

Central gray

Settling tegmental neurons

Bed nucleus of the stria terminalis

Forel's fields

Sojourning and migrating thalamic neurons

Substantia innominata

Migrating basal telencephalic neurons

Lateral preoptic/hypothalamic areas

Medial forebrain bundle

Substantia nigra

Mesencephalic (tegmental) NEP

Mesencephalic (tectal, inferior colliculus) NEP

Migratory stream at base of telencephalon

Medial forebrain bundle

Inter-peduncular nucleus

Trochlear nucleus (IV)

Migrating preoptic neurons

Mammillary body

Nerve IV (trochlear)

Preoptic NEP

Ventro-medial nucleus

Isthmal NEP?

Superior cerebellar peduncle

Optic tract and chiasm

Pituitary gland

Medial longitudinal fasciculus

Reticular

Pontine NEP

Sojourning and migrating deep neurons

External germinal layer

Anterior extramural migratory stream (pontine gray and reticular tegmental neurons)

Pontomedullary trench

Sprouting fibers of hook bundle?

Dorsal rhombic lip

Premigratory facial motor nucleus (VII) neurons?

formation

Prepositus nucleus

Sojourning and migrating Purkinje cells

Choroid plexus stem cells

Subpial GEP?

Upper medullary NEP

Cerebellar NEP

Medial lemniscus?

Inferior olive fibrous capsule

Medullary velum

Inferior olive

Lower medullary NEP

Ventral rhombic lip

FONT KEY:
Germinal zone - Helvetica bold
Transient structure - Times bold italic
Permanent structure - Times Roman or **Bold**

Posterior intramural migratory stream (inferior olive neurons)

Migrating gracile nucleus neurons?

Posterior extramural migratory stream (external cuneate and lateral reticular neurons)

Gracile fasciculus?

Migrating spinal dorsal gray neurons

Midline glial raphe system (structural support for brainstem flexures)

Spinal GEP

Arrows indicate the presumed *direction of neuron migration* from neuroepithelial sources.

Migrating interfascicular glia infiltrate midline raphe glia?

Spinal NEP

ABBREVIATIONS:
GEP - Glioepithelium
NEP - Neuroepithelium
STF - Stratified transitional field
SVZ - Subventricular zone

Ventral funiculus

Central autonomic area

Ventral gray matter

Dorsal funiculus

Intermediate gray matter

**HEAD STRUCTURES,
MAJOR BRAIN REGIONS,
AND VENTRICULAR
DIVISIONS**

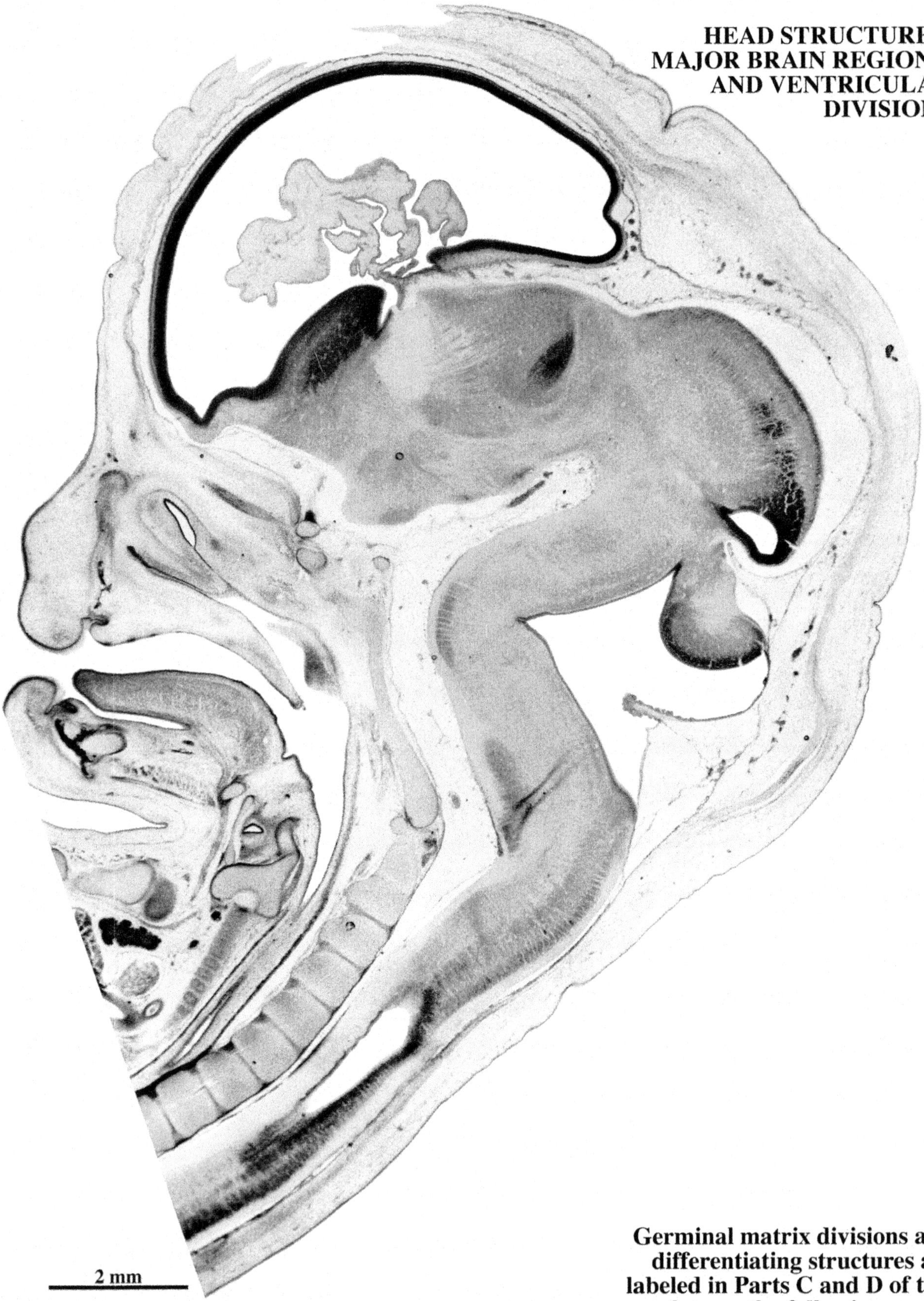

2 mm

**Germinal matrix divisions and
differentiating structures are
labeled in Parts C and D of this
plate on the following pages.**

Dashed lines indicate staining
and/or sectioning artifacts.

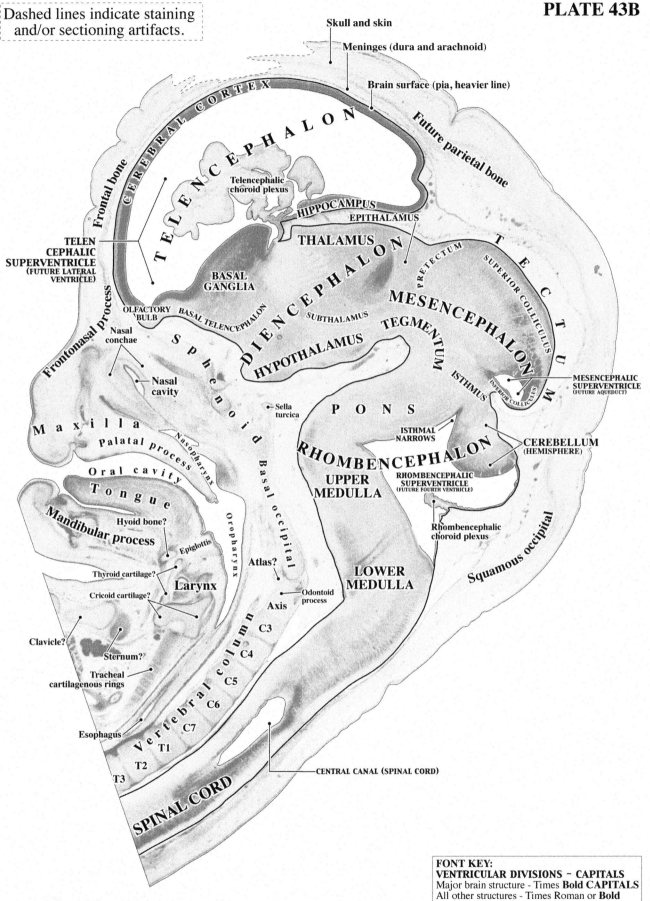

Skull and skin

Meninges (dura and arachnoid)

Brain surface (pia, heavier line)

Future parietal bone

CEREBRAL CORTEX

TELENCEPHALON

Frontal bone

Telencephalic
choroid plexus

HIPPOCAMPUS

EPITHALAMUS

THALAMUS

TELEN
CEPHALIC
SUPERVENTRICLE
(FUTURE LATERAL
VENTRICLE)

BASAL
GANGLIA

DIENCEPHALON

PRETECTUM

SUPERIOR COLLICULUS

TECTUM

MESENCEPHALON

OLFACTORY
BULB

BASAL TELENCEPHALON

SUBTHALAMUS

HYPOTHALAMUS

TEGMENTUM

ISTHMUS

INFERIOR COLLICULUS

MESENCEPHALIC
SUPERVENTRICLE
(FUTURE AQUEDUCT)

Frontonasal process

Nasal
conchae

Sphenoid

Nasal
cavity

Sella
turcica

PONS

Maxilla

Palatal process

Nasopharynx

Basal occipital

RHOMBENCEPHALON

ISTHMAL
NARROWS

CEREBELLUM
(HEMISPHERE)

Oral cavity

Tongue

UPPER
MEDULLA

RHOMBENCEPHALIC
SUPERVENTRICLE
(FUTURE FOURTH VENTRICLE)

Mandibular process

Hyoid bone?

Oropharynx

Rhombencephalic
choroid plexus

Epiglottis

Thyroid cartilage?

Larynx

Atlas?

LOWER
MEDULLA

Squamous occipital

Cricoid cartilage?

Axis

Odontoid
process

Clavicle?

Sternum?

C3

Tracheal
cartilagenous rings

C4

C5

C6

Esophagus

Vertebral column

C7

T1

T2

T3

CENTRAL CANAL (SPINAL CORD)

SPINAL CORD

FONT KEY:
VENTRICULAR DIVISIONS – CAPITALS
Major brain structure - Times **Bold CAPITALS**
All other structures - Times Roman or **Bold**

PLATE 43C

GERMINAL MATRIX DIVI-
SIONS AND DIFFEREN-
TIATING BRAIN
STRUCTURES

See high-magnification
views of the diencephalon
and basal telencephalon in
Plates 53A and B.

BRAINSTEM FLEXURES

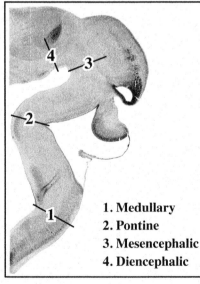

1. Medullary
2. Pontine
3. Mesencephalic
4. Diencephalic

The head, major brain
structures, and
ventricular divisions
are labeled in Parts A
and B of this plate on
the preceding pages.

Left side — Midline
Right side

2 mm

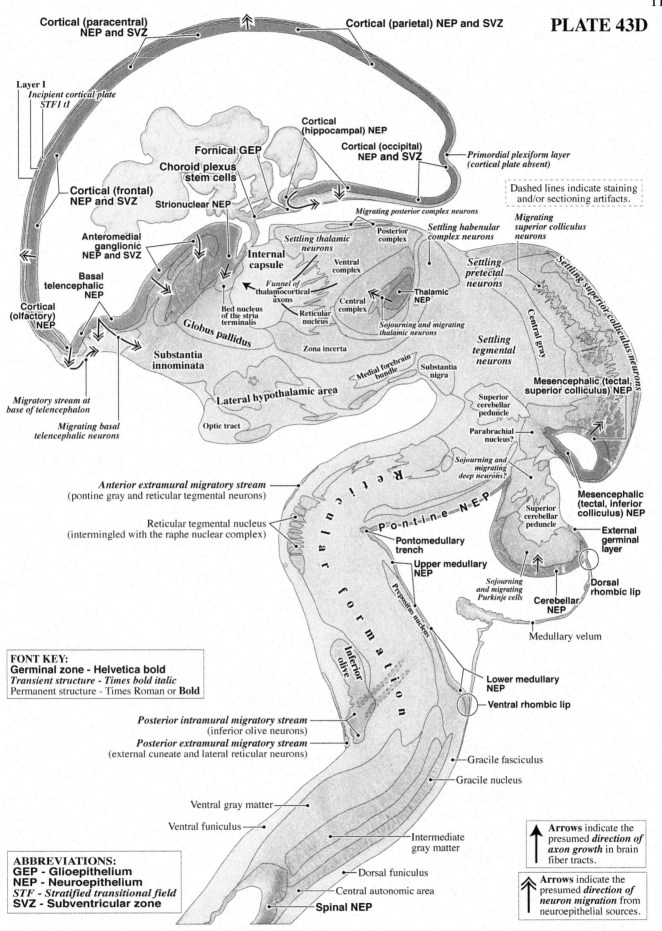

Cortical (paracentral) NEP and SVZ

Cortical (parietal) NEP and SVZ

Layer I
Incipient cortical plate
STF1 t1

Cortical (frontal) NEP and SVZ

Fornical GEP

Choroid plexus stem cells

Strionuclear NEP

Cortical (hippocampal) NEP

Cortical (occipital) NEP and SVZ

Primordial plexiform layer (cortical plate absent)

Migrating posterior complex neurons

Migrating superior colliculus neurons

Dashed lines indicate staining and/or sectioning artifacts.

Anteromedial ganglionic NEP and SVZ

Settling thalamic neurons

Posterior complex

Settling habenular complex neurons

Settling pretectal neurons

Settling superior colliculus neurons

Basal telencephalic NEP

Internal capsule

Ventral complex

Settling tegmental neurons

Cortical (olfactory) NEP

Funnel of thalamocortical axons

Central complex

Thalamic NEP

Central gray

Reticular nucleus

Bed nucleus of the stria terminalis

Globus pallidus

Sojourning and migrating thalamic neurons

Substantia innominata

Zona incerta

Substantia nigra

Mesencephalic (tectal, superior colliculus) NEP

Migratory stream at base of telencephalon

Lateral hypothalamic area

Medial forebrain bundle

Superior cerebellar peduncle

Migrating basal telencephalic neurons

Optic tract

Parabrachial nucleus?

Sojourning and migrating deep neurons?

Mesencephalic (tectal, inferior colliculus) NEP

Reticular formation

Pontine NEP

Superior cerebellar peduncle

External germinal layer

Anterior extramural migratory stream
(pontine gray and reticular tegmental neurons)

Reticular tegmental nucleus
(intermingled with the raphe nuclear complex)

Pontomedullary trench

Dorsal rhombic lip

Upper medullary NEP

Sojourning and migrating Purkinje cells

Cerebellar NEP

Prepositus nucleus

Medullary velum

Inferior olive

Lower medullary NEP

FONT KEY:
Germinal zone - Helvetica bold
Transient structure - Times bold italic
Permanent structure - Times Roman or **Bold**

Ventral rhombic lip

Posterior intramural migratory stream
(inferior olive neurons)

Posterior extramural migratory stream
(external cuneate and lateral reticular neurons)

Gracile fasciculus

Gracile nucleus

Ventral gray matter

Ventral funiculus

Intermediate gray matter

ABBREVIATIONS:
GEP - Glioepithelium
NEP - Neuroepithelium
STF - Stratified transitional field
SVZ - Subventricular zone

Dorsal funiculus

Central autonomic area

Spinal NEP

Arrows indicate the presumed *direction of axon growth* in brain fiber tracts.

Arrows indicate the presumed *direction of neuron migration* from neuroepithelial sources.

PLATE 44A

CR 33 mm, GW 9.6, C145
Sagittal, Slide 16, Section 2
Left side of brain

**HEAD STRUCTURES,
MAJOR BRAIN REGIONS,
AND VENTRICULAR
DIVISIONS**

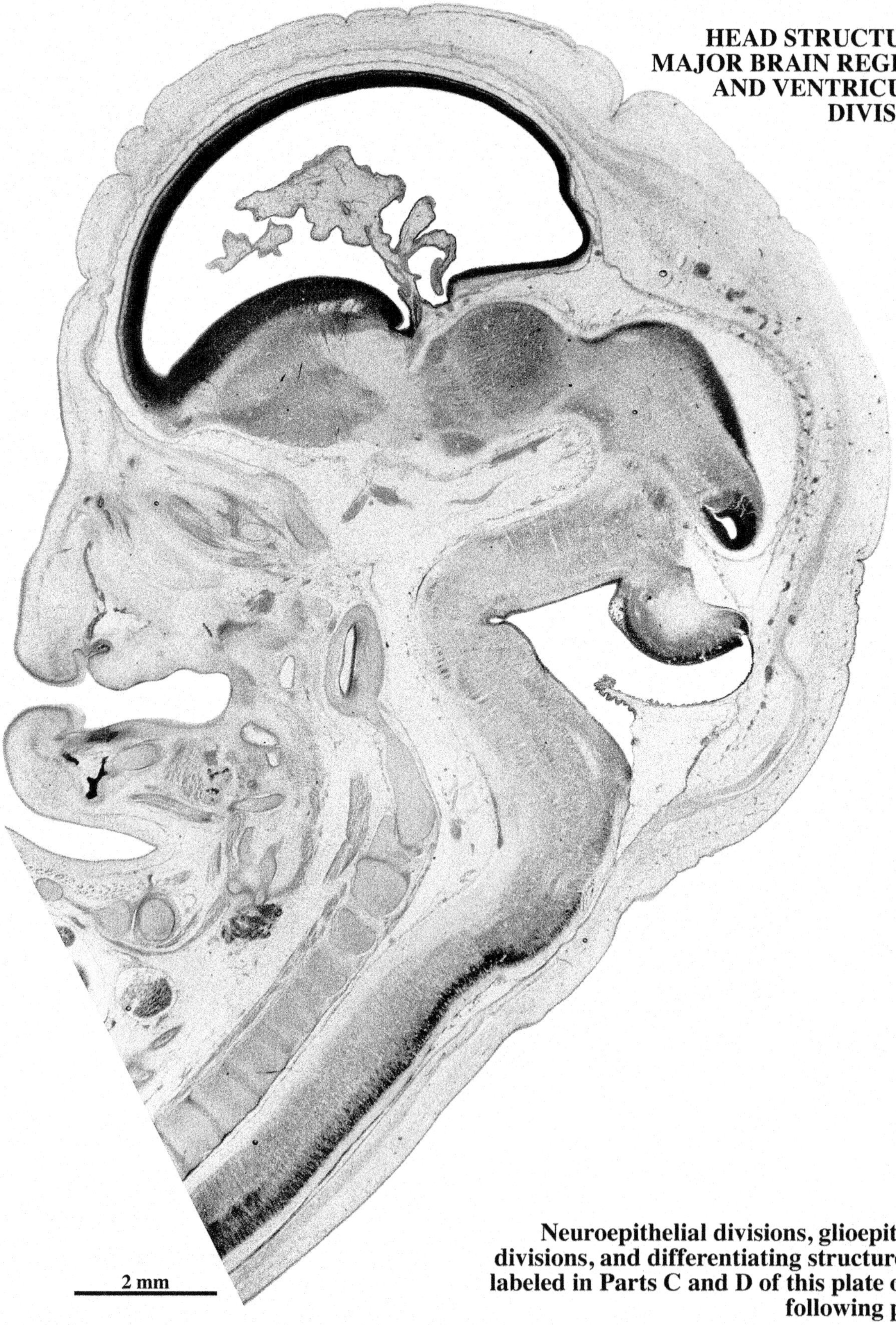

2 mm

Neuroepithelial divisions, glioepithelial
divisions, and differentiating structures are
labeled in Parts C and D of this plate on the
following pages.

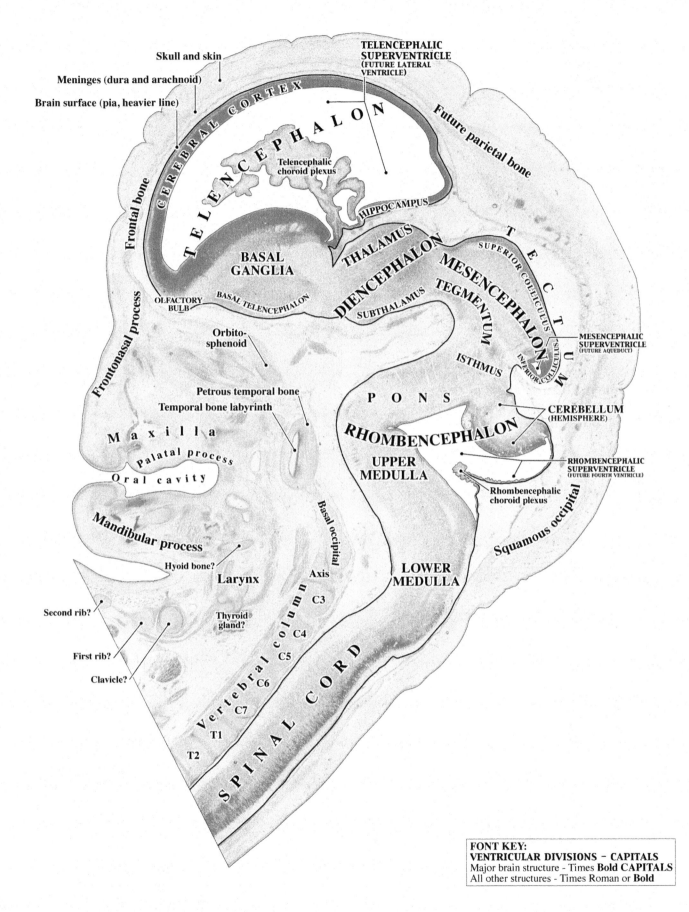

Skull and skin

Meninges (dura and arachnoid)

Brain surface (pia, heavier line)

TELENCEPHALIC SUPERVENTRICLE
(FUTURE LATERAL VENTRICLE)

CEREBRAL CORTEX

Future parietal bone

Frontal bone

TELENCEPHALON

Telencephalic choroid plexus

HIPPOCAMPUS

THALAMUS

T E C T U M

SUPERIOR COLLICULUS

BASAL GANGLIA

DIENCEPHALON

MESENCEPHALON

TEGMENTUM

OLFACTORY BULB

BASAL TELENCEPHALON

SUBTHALAMUS

MESENCEPHALIC SUPERVENTRICLE
(FUTURE AQUEDUCT)

INFERIOR COLLICULUS

Frontonasal process

Orbito-sphenoid

ISTHMUS

Petrous temporal bone

Temporal bone labyrinth

P O N S

CEREBELLUM
(HEMISPHERE)

M a x i l l a

RHOMBENCEPHALON

Palatal process

UPPER MEDULLA

Oral cavity

RHOMBENCEPHALIC SUPERVENTRICLE
(FUTURE FOURTH VENTRICLE)

Mandibular process

Rhombencephalic choroid plexus

Hyoid bone?

Basal occipital

Squamous occipital

Larynx

Axis

Second rib?

C3

Thyroid gland?

C4

First rib?

C5

Vertebral column

C6

Clavicle?

C7

T1

LOWER MEDULLA

T2

SPINAL CORD

FONT KEY:
VENTRICULAR DIVISIONS – CAPITALS
Major brain structure - Times **Bold CAPITALS**
All other structures - Times Roman or **Bold**

122

**GERMINAL MATRIX
DIVISIONS AND
DIFFERENTIATING
BRAIN STRUCTURES**

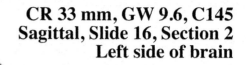

**See high-magnification views of the
diencephalon and basal telencephalon from
Slide 17 Section 2 in Plates 54A and B.**

BRAINSTEM FLEXURES

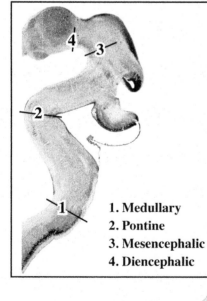

1. Medullary
2. Pontine
3. Mesencephalic
4. Diencephalic

The head, major brain
structures, and
ventricular divisions
are labeled in Parts A
and B of this plate on
the preceding pages.

2 mm

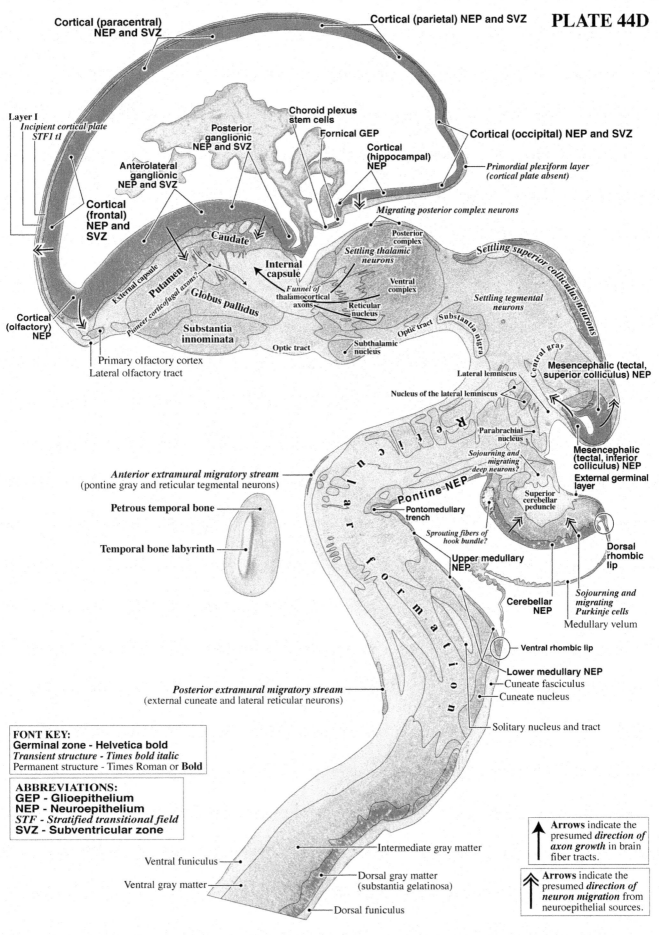

Cortical (paracentral)
NEP and SVZ

Cortical (parietal) NEP and SVZ

Cortical (occipital) NEP and SVZ

*Primordial plexiform layer
(cortical plate absent)*

Choroid plexus
stem cells

Posterior
ganglionic
NEP and SVZ

Fornical GEP

Cortical
(hippocampal)
NEP

Layer I
*Incipient cortical plate
STF1 t1*

Anterolateral
ganglionic
NEP and SVZ

Cortical
(frontal)
NEP and
SVZ

Migrating posterior complex neurons

Posterior
complex

*Settling thalamic
neurons*

Settling superior colliculus neurons

Caudate

**Internal
capsule**

Ventral
complex

*Settling tegmental
neurons*

External capsule

Putamen

*Funnel of
thalamocortical
axons*

Reticular
nucleus

Globus pallidus

Pioneer corticofugal axons?

Cortical
(olfactory)
NEP

Substantia
innominata

Optic tract

Subthalamic
nucleus

Substantia nigra

Central gray

**Mesencephalic (tectal,
superior colliculus) NEP**

Primary olfactory cortex
Lateral olfactory tract

Optic tract

Lateral lemniscus

Nucleus of the lateral lemniscus

R e t i c u l a r   f o r m a t i o n

Parabrachial
nucleus

*Sojourning and
migrating
deep neurons?*

**Mesencephalic
(tectal, inferior
colliculus) NEP**

**External germinal
layer**

*Anterior extramural migratory stream
(pontine gray and reticular tegmental neurons)*

Petrous temporal bone

Temporal bone labyrinth

Pontine NEP

Pontomedullary
trench

*Sprouting fibers of
hook bundle?*

Superior
cerebellar
peduncle

**Dorsal
rhombic
lip**

**Upper medullary
NEP**

**Cerebellar
NEP**

*Sojourning and
migrating
Purkinje cells*

Medullary velum

Ventral rhombic lip

Lower medullary NEP
Cuneate fasciculus
Cuneate nucleus

*Posterior extramural migratory stream
(external cuneate and lateral reticular neurons)*

Solitary nucleus and tract

FONT KEY:
Germinal zone - Helvetica bold
Transient structure - Times bold italic
Permanent structure - Times Roman or **Bold**

ABBREVIATIONS:
GEP - Glioepithelium
NEP - Neuroepithelium
STF - Stratified transitional field
SVZ - Subventricular zone

Intermediate gray matter

Ventral funiculus

Dorsal gray matter
(substantia gelatinosa)

Ventral gray matter

Dorsal funiculus

Arrows indicate the
presumed *direction of
axon growth* in brain
fiber tracts.

Arrows indicate the
presumed *direction of
neuron migration* from
neuroepithelial sources.

PLATE 45A

**HEAD STRUCTURES,
MAJOR BRAIN REGIONS,
AND VENTRICULAR
DIVISIONS**

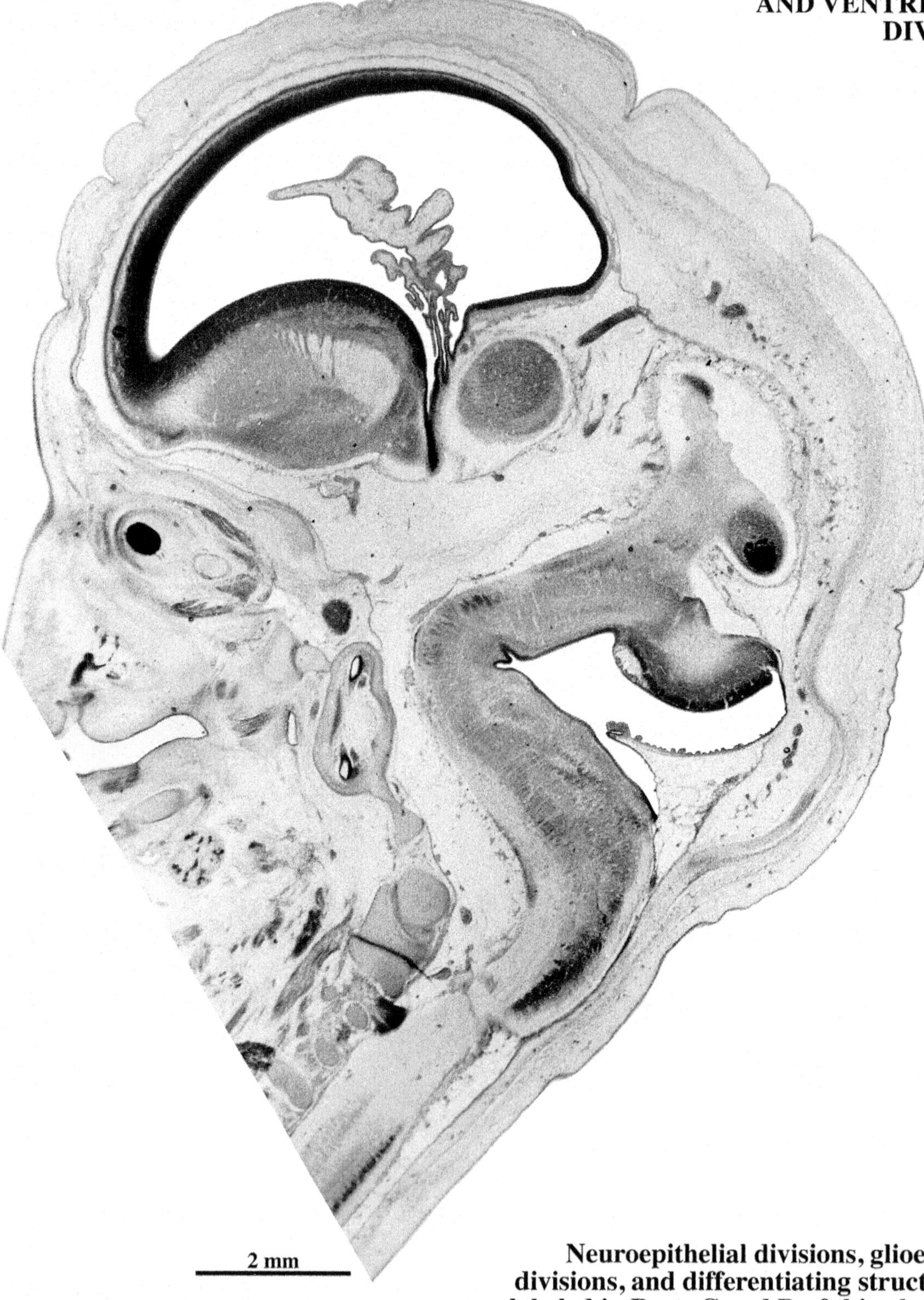

2 mm

Neuroepithelial divisions, glioepithelial divisions, and differentiating structures are labeled in Parts C and D of this plate on the following pages.

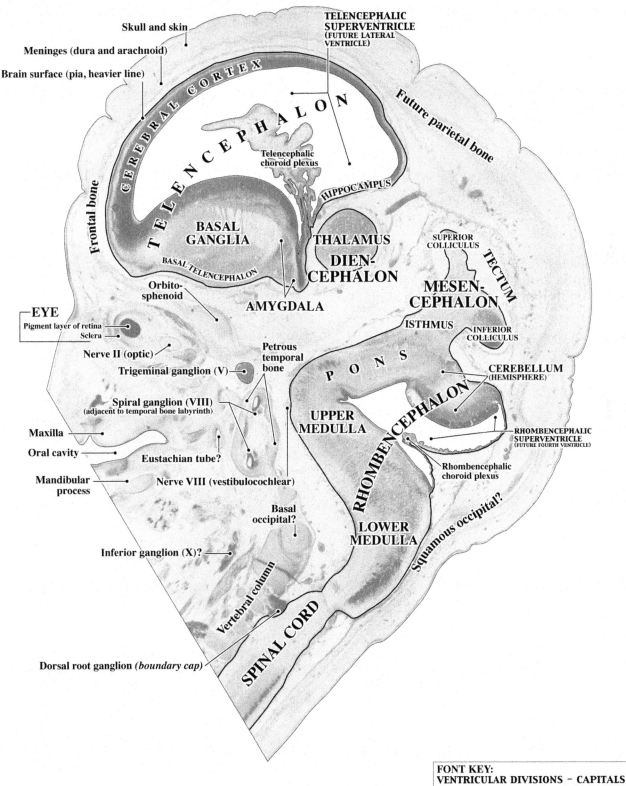

Skull and skin

Meninges (dura and arachnoid)

Brain surface (pia, heavier line)

TELENCEPHALIC SUPERVENTRICLE (FUTURE LATERAL VENTRICLE)

CEREBRAL CORTEX

TELENCEPHALON

Future parietal bone

Telencephalic choroid plexus

HIPPOCAMPUS

Frontal bone

BASAL GANGLIA

BASAL TELENCEPHALON

THALAMUS

DIEN-CEPHALON

SUPERIOR COLLICULUS

TECTUM

Orbito-sphenoid

AMYGDALA

MESEN-CEPHALON

ISTHMUS

INFERIOR COLLICULUS

EYE

Pigment layer of retina

Sclera

Nerve II (optic)

Trigeminal ganglion (V)

Petrous temporal bone

P O N S

CEREBELLUM (HEMISPHERE)

Spiral ganglion (VIII) (adjacent to temporal bone labyrinth)

UPPER MEDULLA

RHOMBENCEPHALON

RHOMBENCEPHALIC SUPERVENTRICLE (FUTURE FOURTH VENTRICLE)

Maxilla

Oral cavity

Eustachian tube?

Mandibular process

Nerve VIII (vestibulocochlear)

Rhombencephalic choroid plexus

Basal occipital?

LOWER MEDULLA

Squamous occipital?

Inferior ganglion (X)?

Vertebral column

Dorsal root ganglion (*boundary cap*)

SPINAL CORD

FONT KEY:
VENTRICULAR DIVISIONS – CAPITALS
Major brain structure - Times **Bold CAPITALS**
All other structures - Times Roman or **Bold**

PLATE 45C

CR 33 mm, GW 9.6, C145
Sagittal, Slide 15, Section 1
Left side of brain

**GERMINAL MATRIX
DIVISIONS AND
DIFFERENTIATING
BRAIN STRUCTURES**

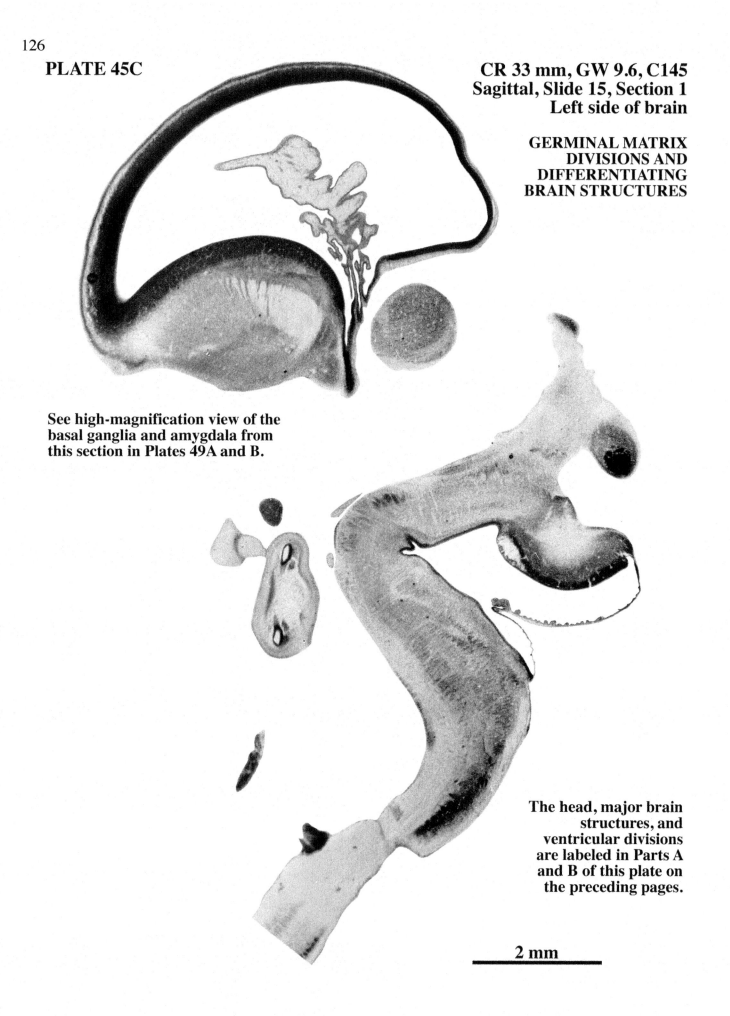

See high-magnification view of the
basal ganglia and amygdala from
this section in Plates 49A and B.

The head, major brain
structures, and
ventricular divisions
are labeled in Parts A
and B of this plate on
the preceding pages.

2 mm

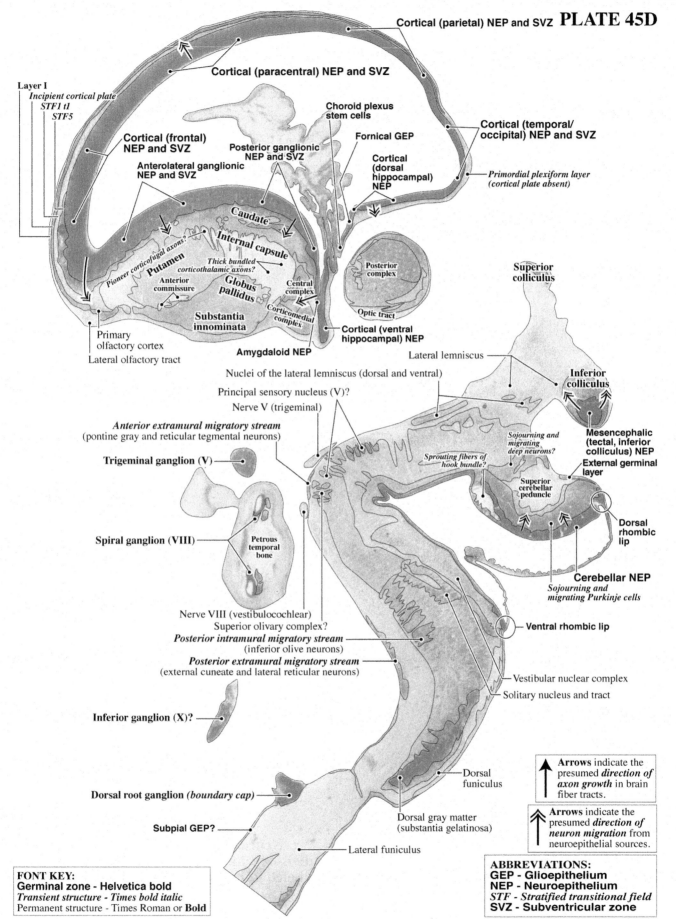

Cortical (parietal) NEP and SVZ **PLATE 45D**

Cortical (paracentral) NEP and SVZ

Layer I
Incipient cortical plate
STF1 t1
STF5

Choroid plexus stem cells

Fornical GEP

Cortical (temporal/occipital) NEP and SVZ

Cortical (frontal) NEP and SVZ

Posterior ganglionic NEP and SVZ

Anterolateral ganglionic NEP and SVZ

Cortical (dorsal hippocampal) NEP

Primordial plexiform layer (cortical plate absent)

Caudate

Internal capsule

Pioneer corticofugal axons?

Thick bundled corticothalamic axons?

Putamen

Anterior commissure

Globus pallidus

Central complex

Posterior complex

Superior colliculus

Substantia innominata

Corticomedial complex

Cortical (ventral hippocampal) NEP

Optic tract

Primary olfactory cortex

Lateral olfactory tract

Amygdaloid NEP

Lateral lemniscus

Inferior colliculus

Nuclei of the lateral lemniscus (dorsal and ventral)

Principal sensory nucleus (V)?

Nerve V (trigeminal)

Anterior extramural migratory stream
(pontine gray and reticular tegmental neurons)

Mesencephalic (tectal, inferior colliculus) NEP

Sojourning and migrating deep neurons?

Sprouting fibers of hook bundle?

External germinal layer

Trigeminal ganglion (V)

Superior cerebellar peduncle

Dorsal rhombic lip

Spiral ganglion (VIII)

Petrous temporal bone

Cerebellar NEP
Sojourning and migrating Purkinje cells

Nerve VIII (vestibulocochlear)

Superior olivary complex?

Ventral rhombic lip

Posterior intramural migratory stream
(inferior olive neurons)

Posterior extramural migratory stream
(external cuneate and lateral reticular neurons)

Vestibular nuclear complex

Solitary nucleus and tract

Inferior ganglion (X)?

Dorsal funiculus

Dorsal root ganglion *(boundary cap)*

Dorsal gray matter (substantia gelatinosa)

Subpial GEP?

Lateral funiculus

Arrows indicate the presumed *direction of axon growth* in brain fiber tracts.

Arrows indicate the presumed *direction of neuron migration* from neuroepithelial sources.

FONT KEY:
Germinal zone - **Helvetica bold**
Transient structure - Times bold italic
Permanent structure - Times Roman or **Bold**

ABBREVIATIONS:
GEP - Glioepithelium
NEP - Neuroepithelium
STF - Stratified transitional field
SVZ - Subventricular zone

PLATE 46A

**HEAD STRUCTURES,
MAJOR BRAIN REGIONS,
AND VENTRICULAR
DIVISIONS**

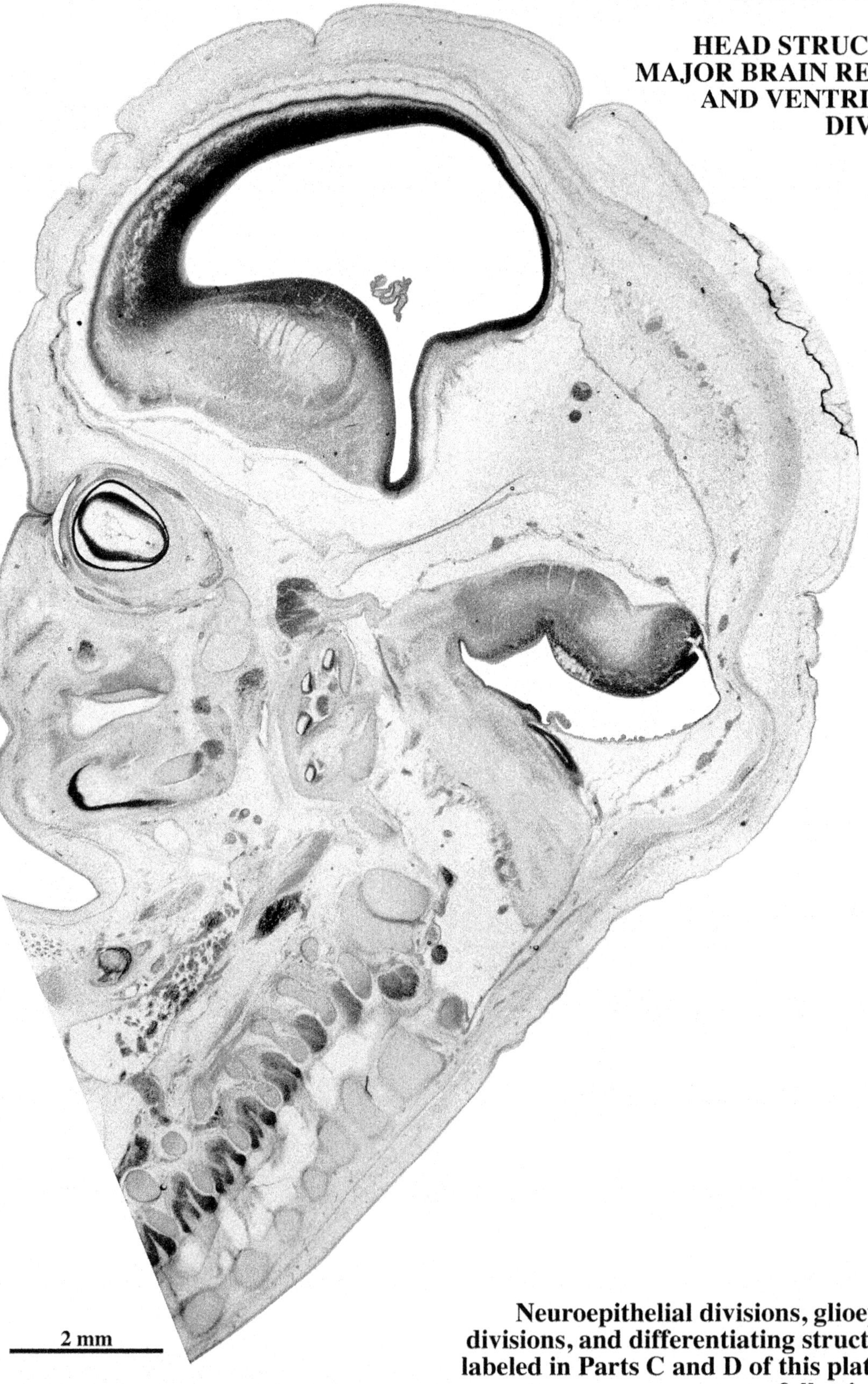

2 mm

Neuroepithelial divisions, glioepithelial
divisions, and differentiating structures are
labeled in Parts C and D of this plate on the
following pages.

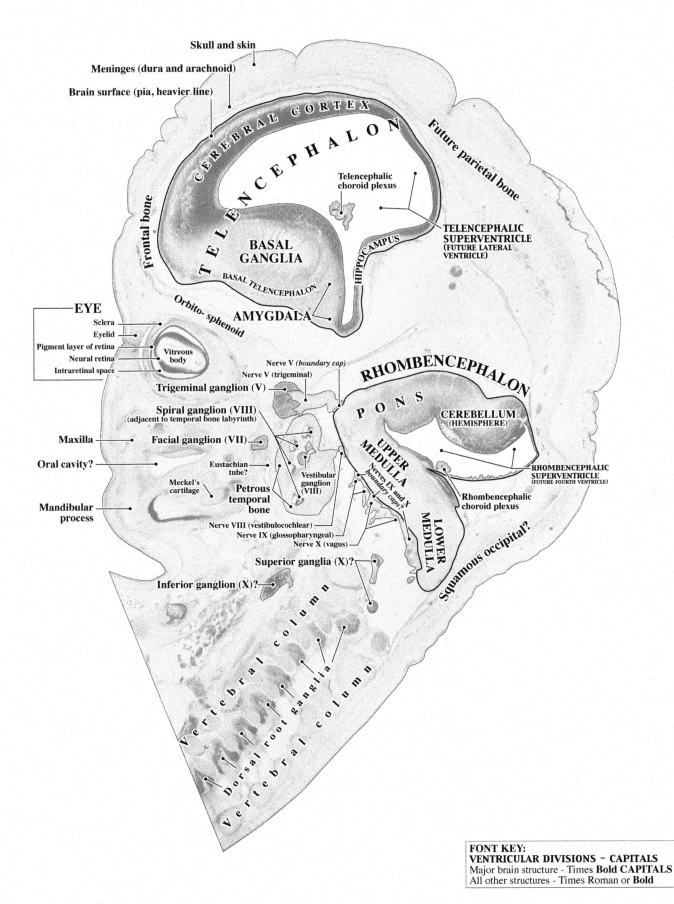

Skull and skin

Meninges (dura and arachnoid)

Brain surface (pia, heavier line)

CEREBRAL CORTEX

TELENCEPHALON

Future parietal bone

Telencephalic
choroid plexus

Frontal bone

BASAL
GANGLIA

BASAL TELENCEPHALON

HIPPOCAMPUS

TELENCEPHALIC
SUPERVENTRICLE
(FUTURE LATERAL
VENTRICLE)

EYE

Sclera

Eyelid

Pigment layer of retina

Neural retina

Intraretinal space

Orbito- sphenoid

AMYGDALA

Vitreous
body

Nerve V (boundary cap)

Nerve V (trigeminal)

RHOMBENCEPHALON

Trigeminal ganglion (V)

PONS

CEREBELLUM
(HEMISPHERE)

Spiral ganglion (VIII)
(adjacent to temporal bone labyrinth)

Maxilla

Facial ganglion (VII)

UPPER
MEDULLA

Oral cavity?

Eustachian
tube?

Vestibular
ganglion
(VIII)

Nerves IX and X
boundary caps?

RHOMBENCEPHALIC
SUPERVENTRICLE
(FUTURE FOURTH VENTRICLE)

Meckel's
cartilage

Petrous
temporal
bone

Rhombencephalic
choroid plexus

Mandibular
process

Nerve VIII (vestibulocochlear)

Nerve IX (glossopharyngeal)

Nerve X (vagus)

LOWER
MEDULLA

Squamous occipital?

Superior ganglia (X)?

Inferior ganglion (X)?

Vertebral column

Dorsal root ganglia

Vertebral column

FONT KEY:
VENTRICULAR DIVISIONS – CAPITALS
Major brain structure - Times **Bold CAPITALS**
All other structures - Times Roman or **Bold**

PLATE 46C

**GERMINAL MATRIX
DIVISIONS AND
DIFFERENTIATING
BRAIN STRUCTURES**

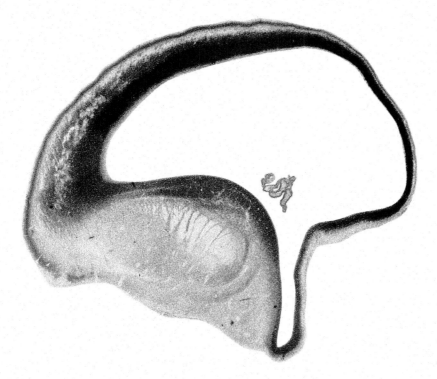

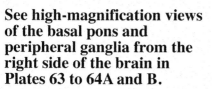

See high-magnification views
of the basal pons and
peripheral ganglia from the
right side of the brain in
Plates 63 to 64A and B.

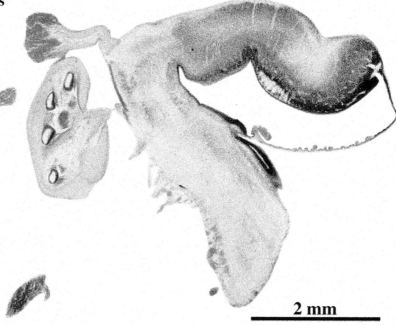

2 mm

The head, major brain structures, and ventricular
divisions are labeled in Parts A and B of this plate
on the preceding pages.

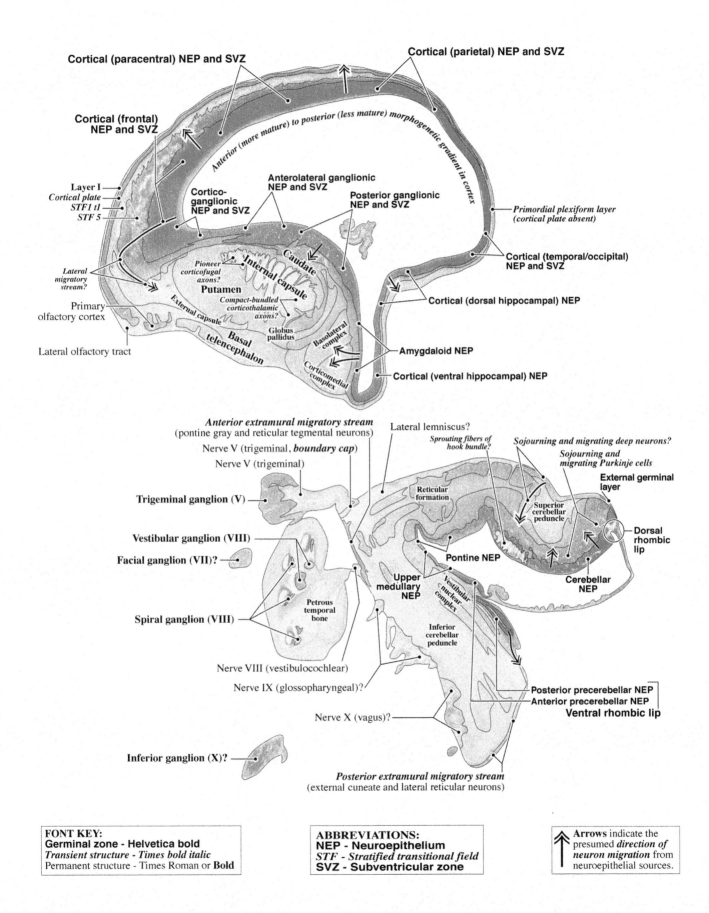

Cortical (paracentral) NEP and SVZ

Cortical (parietal) NEP and SVZ

Cortical (frontal) NEP and SVZ

Anterior (more mature) to posterior (less mature) morphogenetic gradient in cortex

Layer I
Cortical plate
STF1 tl
STF 5

Anterolateral ganglionic NEP and SVZ

Cortico-ganglionic NEP and SVZ

Posterior ganglionic NEP and SVZ

Primordial plexiform layer (cortical plate absent)

Pioneer corticofugal axons?

Caudate

Internal capsule

Lateral migratory stream?

Putamen

External capsule

Compact-bundled corticothalamic axons?

Globus pallidus

Basolateral complex

Cortical (temporal/occipital) NEP and SVZ

Cortical (dorsal hippocampal) NEP

Primary olfactory cortex

Basal telencephalon

Lateral olfactory tract

Corticomedial complex

Amygdaloid NEP

Cortical (ventral hippocampal) NEP

Anterior extramural migratory stream
(pontine gray and reticular tegmental neurons)

Nerve V (trigeminal, *boundary cap*)

Nerve V (trigeminal)

Trigeminal ganglion (V)

Lateral lemniscus?

Sprouting fibers of hook bundle?

Reticular formation

Sojourning and migrating deep neurons?

Sojourning and migrating Purkinje cells

External germinal layer

Superior cerebellar peduncle

Dorsal rhombic lip

Vestibular ganglion (VIII)

Facial ganglion (VII)?

Petrous temporal bone

Pontine NEP

Cerebellar NEP

Upper medullary NEP

Vestibular nuclear complex

Spiral ganglion (VIII)

Inferior cerebellar peduncle

Nerve VIII (vestibulocochlear)

Nerve IX (glossopharyngeal)?

Posterior precerebellar NEP
Anterior precerebellar NEP
Ventral rhombic lip

Nerve X (vagus)?

Inferior ganglion (X)?

Posterior extramural migratory stream
(external cuneate and lateral reticular neurons)

FONT KEY:
Germinal zone - Helvetica bold
Transient structure - Times bold italic
Permanent structure - Times Roman or **Bold**

ABBREVIATIONS:
NEP - Neuroepithelium
STF - Stratified transitional field
SVZ - Subventricular zone

Arrows indicate the presumed *direction of neuron migration* from neuroepithelial sources.

PLATE 47A

CR 33 mm, GW 9.6, C145
Sagittal, Slide 12, Section 4
Left side of brain

**HEAD STRUCTURES,
MAJOR BRAIN REGIONS,
AND VENTRICULAR
DIVISIONS**

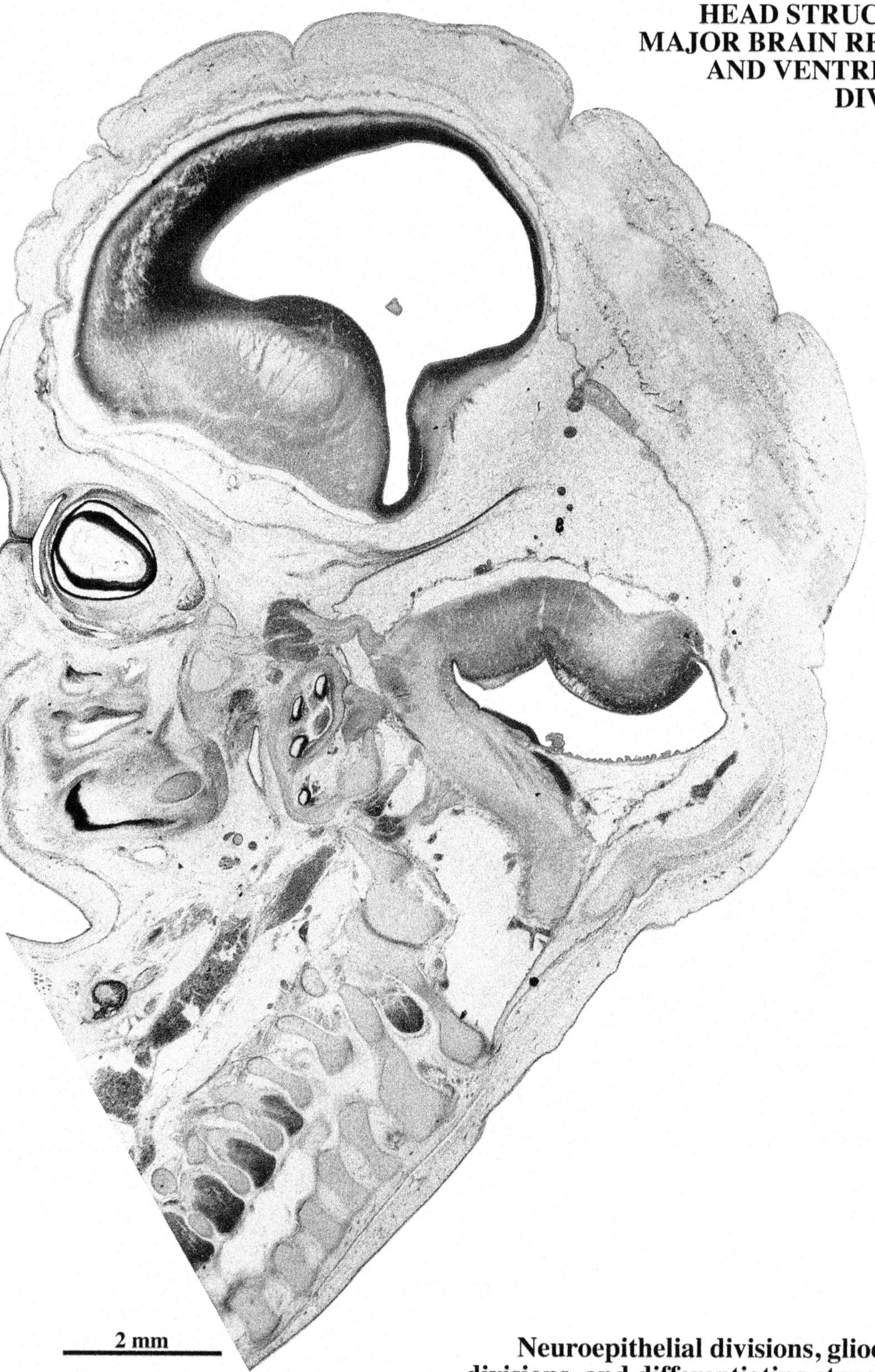

2 mm

Neuroepithelial divisions, glioepithelial
divisions, and differentiating structures are
labeled in Parts C and D of this plate on the
following pages.

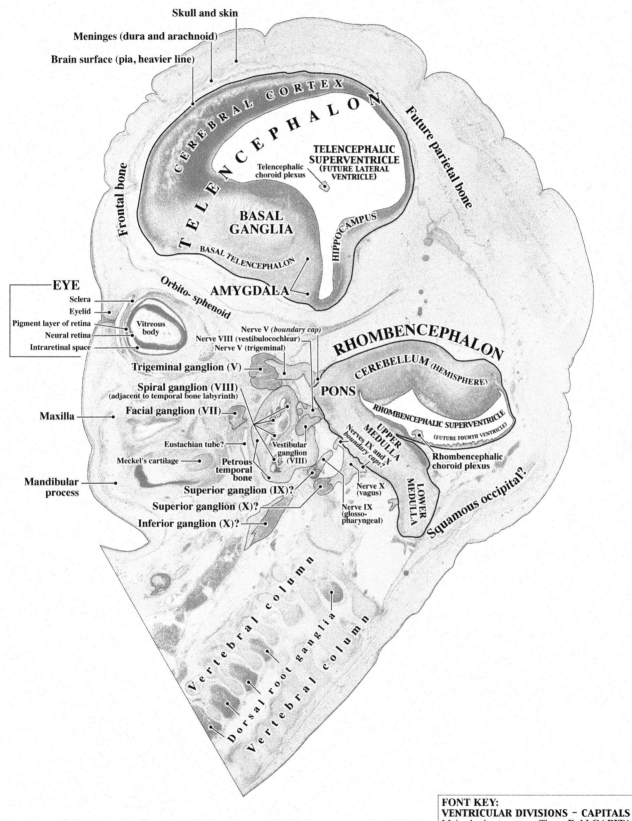

Skull and skin

Meninges (dura and arachnoid)

Brain surface (pia, heavier line)

CEREBRAL CORTEX

TELENCEPHALON

Future parietal bone

Frontal bone

TELENCEPHALIC
SUPERVENTRICLE
(FUTURE LATERAL
VENTRICLE)

Telencephalic
choroid plexus

BASAL
GANGLIA

HIPPOCAMPUS

BASAL TELENCEPHALON

EYE
Sclera
Eyelid
Pigment layer of retina
Neural retina
Intraretinal space

Orbito- sphenoid

AMYGDALA

Vitreous
body

Nerve V (boundary cap)
Nerve VIII (vestibulocochlear)
Nerve V (trigeminal)

RHOMBENCEPHALON

CEREBELLUM (HEMISPHERE)

Trigeminal ganglion (V)

PONS

Spiral ganglion (VIII)
(adjacent to temporal bone labyrinth)

Facial ganglion (VII)

Maxilla

RHOMBENCEPHALIC SUPERVENTRICLE
(FUTURE FOURTH VENTRICLE)

UPPER
MEDULLA

Nerves IX and X
boundary caps

Eustachian tube?

Vestibular
ganglion
(VIII)

Rhombencephalic
choroid plexus

Meckel's cartilage

Petrous
temporal
bone

LOWER
MEDULLA

Mandibular
process

Nerve X
(vagus)

Superior ganglion (IX)?

Superior ganglion (X)?

Nerve IX
(glosso-
pharyngeal)

Squamous occipital?

Inferior ganglion (X)?

Vertebral column

Dorsal root ganglia

Vertebral column

FONT KEY:
VENTRICULAR DIVISIONS – CAPITALS
Major brain structure - Times **Bold CAPITALS**
All other structures - Times Roman or **Bold**

PLATE 47C

**GERMINAL MATRIX DIVISIONS
AND DIFFERENTIATING
BRAIN STRUCTURES**

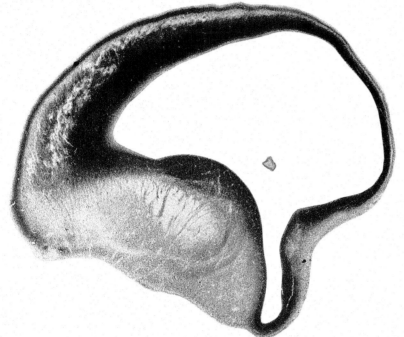

See high-magnification
views of the basal pons
and peripheral ganglia
from the right side of the
brain in Plates 63 to 64A
and B, of the cerebellum
from this Section in Plates
58A and B.

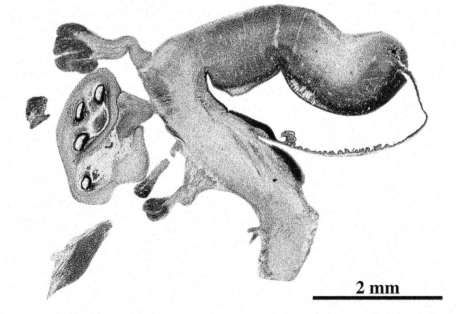

2 mm

The head, major brain structures, and ventricular
divisions are labeled in Parts A and B of this plate
on the preceding pages.

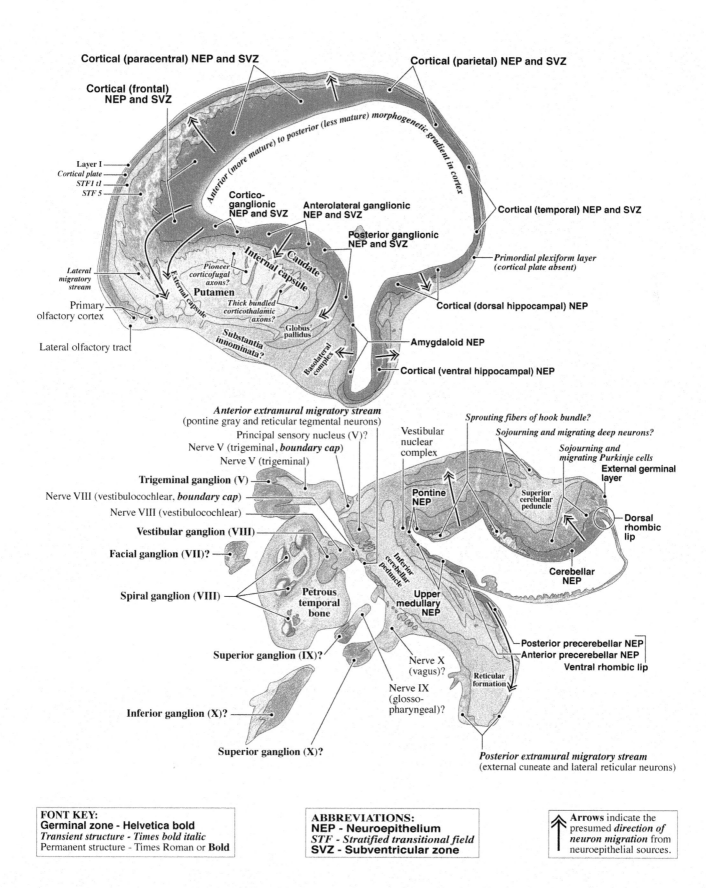

Cortical (paracentral) NEP and SVZ

Cortical (parietal) NEP and SVZ

Cortical (frontal) NEP and SVZ

Anterior (more mature) to posterior (less mature) morphogenetic gradient in cortex

Layer I
Cortical plate
STF1 t1
STF 5

Cortico-ganglionic NEP and SVZ

Anterolateral ganglionic NEP and SVZ

Posterior ganglionic NEP and SVZ

Cortical (temporal) NEP and SVZ

Primordial plexiform layer (cortical plate absent)

Internal capsule

Caudate

Lateral migratory stream

Pioneer corticofugal axons?

External capsule

Putamen

Thick bundled corticothalamic axons?

Globus pallidus

Cortical (dorsal hippocampal) NEP

Primary olfactory cortex

Lateral olfactory tract

Substantia innominata?

Basolateral complex

Amygdaloid NEP

Cortical (ventral hippocampal) NEP

Anterior extramural migratory stream
(pontine gray and reticular tegmental neurons)

Principal sensory nucleus (V)?

Nerve V (trigeminal, *boundary cap*)

Nerve V (trigeminal)

Trigeminal ganglion (V)

Nerve VIII (vestibulocochlear, *boundary cap*)

Nerve VIII (vestibulocochlear)

Vestibular ganglion (VIII)

Facial ganglion (VII)?

Spiral ganglion (VIII)

Vestibular nuclear complex

Sprouting fibers of hook bundle?

Sojourning and migrating deep neurons?

Sojourning and migrating Purkinje cells

External germinal layer

Pontine NEP

Superior cerebellar peduncle

Dorsal rhombic lip

Inferior cerebellar peduncle

Cerebellar NEP

Upper medullary NEP

Petrous temporal bone

Superior ganglion (IX)?

Nerve X (vagus)?

Nerve IX (glosso-pharyngeal)?

Reticular formation

Posterior precerebellar NEP
Anterior precerebellar NEP
Ventral rhombic lip

Inferior ganglion (X)?

Superior ganglion (X)?

Posterior extramural migratory stream
(external cuneate and lateral reticular neurons)

FONT KEY:
Germinal zone - Helvetica bold
Transient structure - Times bold italic
Permanent structure - Times Roman or **Bold**

ABBREVIATIONS:
NEP - Neuroepithelium
STF - Stratified transitional field
SVZ - Subventricular zone

Arrows indicate the presumed *direction of neuron migration* from neuroepithelial sources.

PLATE 48A

**HEAD STRUCTURES,
MAJOR BRAIN REGIONS,
AND VENTRICULAR
DIVISIONS**

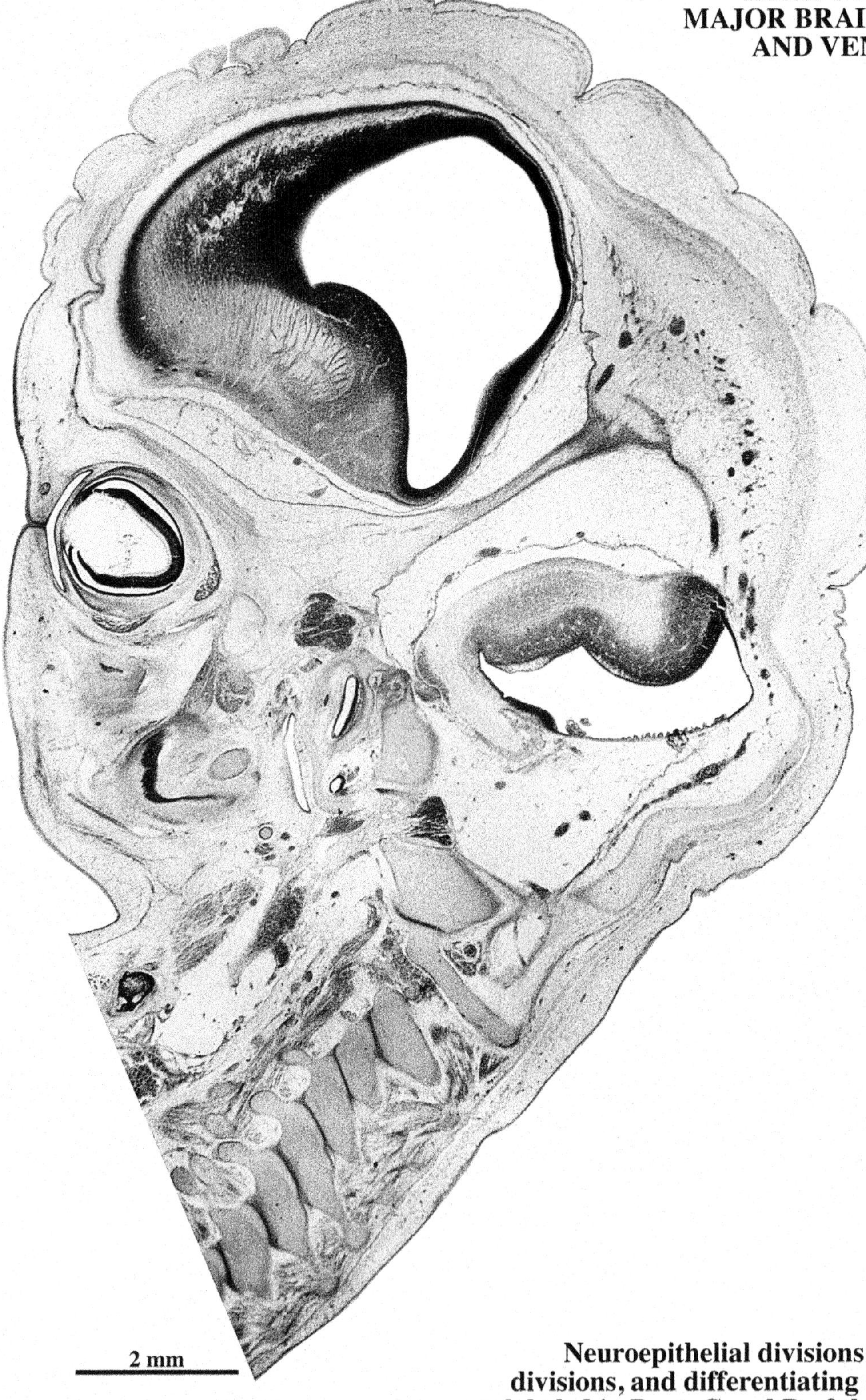

2 mm

Neuroepithelial divisions, glioepithelial
divisions, and differentiating structures are
labeled in Parts C and D of this plate on the
following pages.

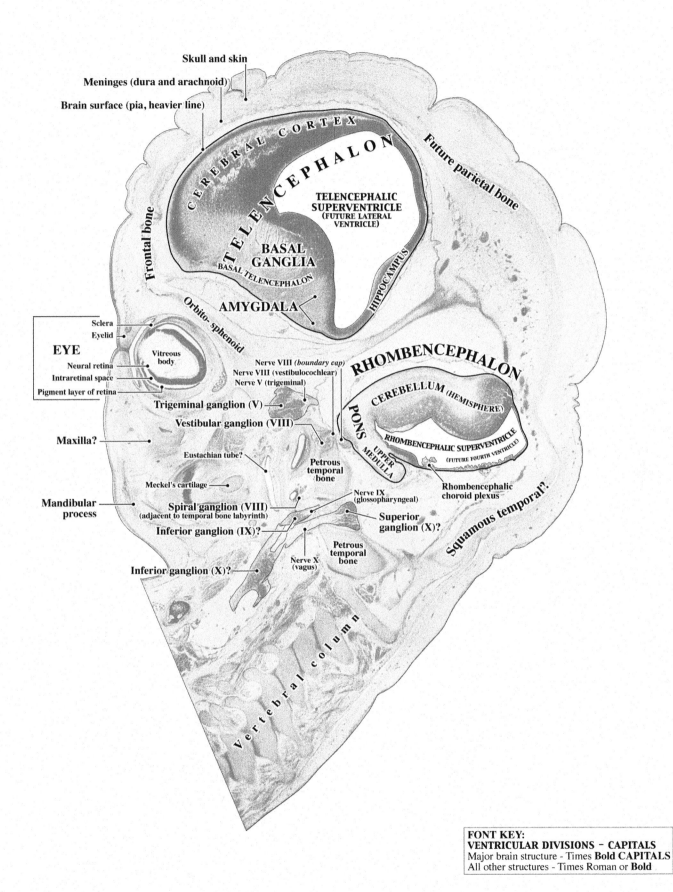

Skull and skin

Meninges (dura and arachnoid)

Brain surface (pia, heavier line)

CEREBRAL CORTEX

TELENCEPHALON

Future parietal bone

TELENCEPHALIC
SUPERVENTRICLE
(FUTURE LATERAL
VENTRICLE)

Frontal bone

BASAL
GANGLIA

BASAL TELENCEPHALON

HIPPOCAMPUS

AMYGDALA

Orbito- sphenoid

Sclera
Eyelid

EYE

Vitreous
body

Neural retina
Intraretinal space

Pigment layer of retina

Nerve VIII *(boundary cap)*
Nerve VIII (vestibulocochlear)
Nerve V (trigeminal)

RHOMBENCEPHALON

CEREBELLUM (HEMISPHERE)

PONS

RHOMBENCEPHALIC SUPERVENTRICLE
(FUTURE FOURTH VENTRICLE)

Trigeminal ganglion (V)

Vestibular ganglion (VIII)

UPPER
MEDULLA

Maxilla?

Eustachian tube?

Petrous
temporal
bone

Rhombencephalic
choroid plexus

Meckel's cartilage

Spiral ganglion (VIII)
(adjacent to temporal bone labyrinth)

Nerve IX
(glossopharyngeal)

Superior
ganglion (X)?

Mandibular
process

Inferior ganglion (IX)?

Inferior ganglion (X)?

Nerve X
(vagus)

Petrous
temporal
bone

Squamous temporal?

Vertebral column

FONT KEY:
VENTRICULAR DIVISIONS - CAPITALS
Major brain structure - Times **Bold CAPITALS**
All other structures - Times Roman or **Bold**

PLATE 48C

CR 33 mm, GW 9.6, C145
Sagittal, Slide 11, Section 4
Left side of brain

**GERMINAL MATRIX DIVISIONS
AND DIFFERENTIATING
BRAIN STRUCTURES**

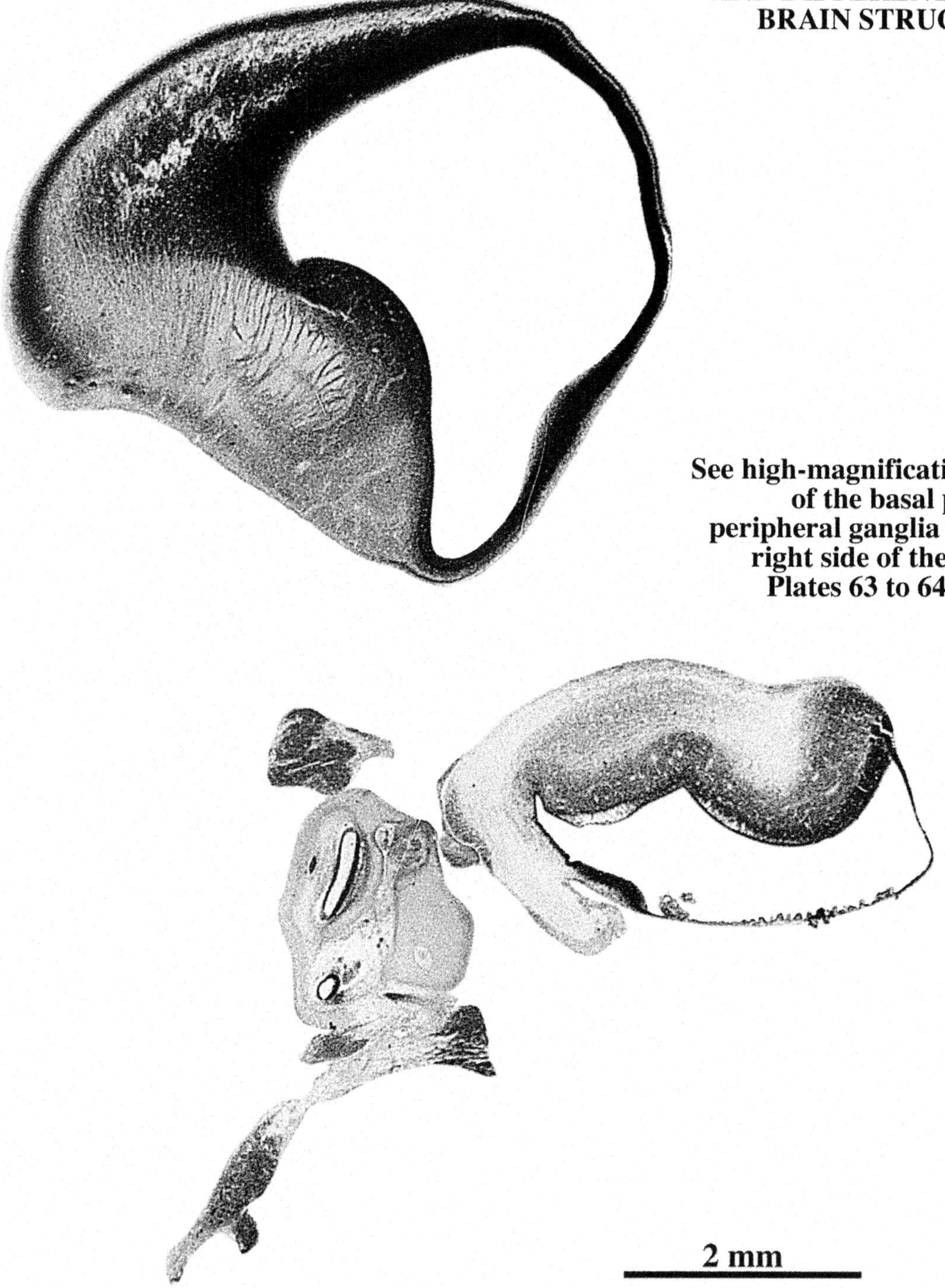

**See high-magnification views
of the basal pons and
peripheral ganglia from the
right side of the brain in
Plates 63 to 64A and B.**

2 mm

**The head, major brain structures, and ventricular
divisions are labeled in Parts A and B of this plate
on the preceding pages.**

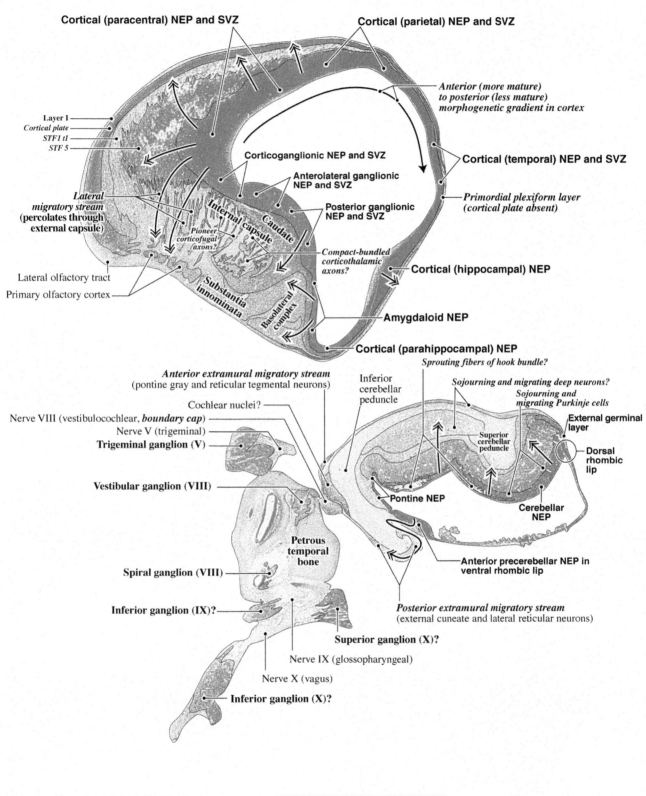

Cortical (paracentral) NEP and SVZ

Cortical (parietal) NEP and SVZ

Anterior (more mature)
to posterior (less mature)
morphogenetic gradient in cortex

Layer I
Cortical plate
STF1 t1
STF 5

Corticoganglionic NEP and SVZ

Anterolateral ganglionic
NEP and SVZ

Posterior ganglionic
NEP and SVZ

Cortical (temporal) NEP and SVZ

Primordial plexiform layer
(cortical plate absent)

Lateral
migratory stream
(percolates through
external capsule)

Internal capsule

Caudate

Pioneer
corticofugal
axons?

Compact-bundled
corticothalamic
axons?

Cortical (hippocampal) NEP

Lateral olfactory tract

Primary olfactory cortex

Substantia
innominata

Basolateral
complex

Amygdaloid NEP

Cortical (parahippocampal) NEP

Sprouting fibers of hook bundle?

Anterior extramural migratory stream
(pontine gray and reticular tegmental neurons)

Inferior
cerebellar
peduncle

Sojourning and migrating deep neurons?

Sojourning and
migrating Purkinje cells

Cochlear nuclei?

Nerve VIII (vestibulocochlear, *boundary cap*)

Nerve V (trigeminal)

Trigeminal ganglion (V)

Superior
cerebellar
peduncle

**External germinal
layer**

**Dorsal
rhombic
lip**

Vestibular ganglion (VIII)

Pontine NEP

Cerebellar
NEP

**Petrous
temporal
bone**

Spiral ganglion (VIII)

Anterior precerebellar NEP in
ventral rhombic lip

Inferior ganglion (IX)?

Superior ganglion (X)?

Posterior extramural migratory stream
(external cuneate and lateral reticular neurons)

Nerve IX (glossopharyngeal)

Nerve X (vagus)

Inferior ganglion (X)?

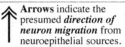

FONT KEY:
Germinal zone - Helvetica bold
Transient structure - Times bold italic
Permanent structure - Times Roman or **Bold**

ABBREVIATIONS:
NEP - Neuroepithelium
STF - Stratified transitional field
SVZ - Subventricular zone

Arrows indicate the
presumed *direction of*
neuron migration from
neuroepithelial sources.

BASAL GANGLIA, BASAL
TELENCEPHALON, AND AMYGDALA

PLATE 49A

CR 33 mm, GW 9.6, C145
Sagittal, Slide 15, Section 1

0.5 mm

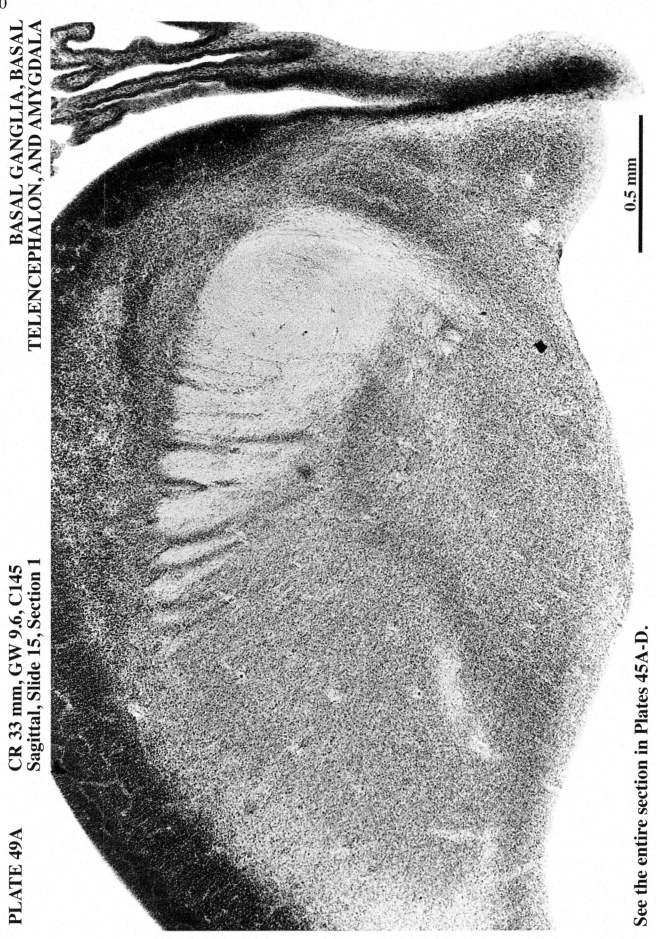

See the entire section in Plates 45A-D.

PLATE 49B

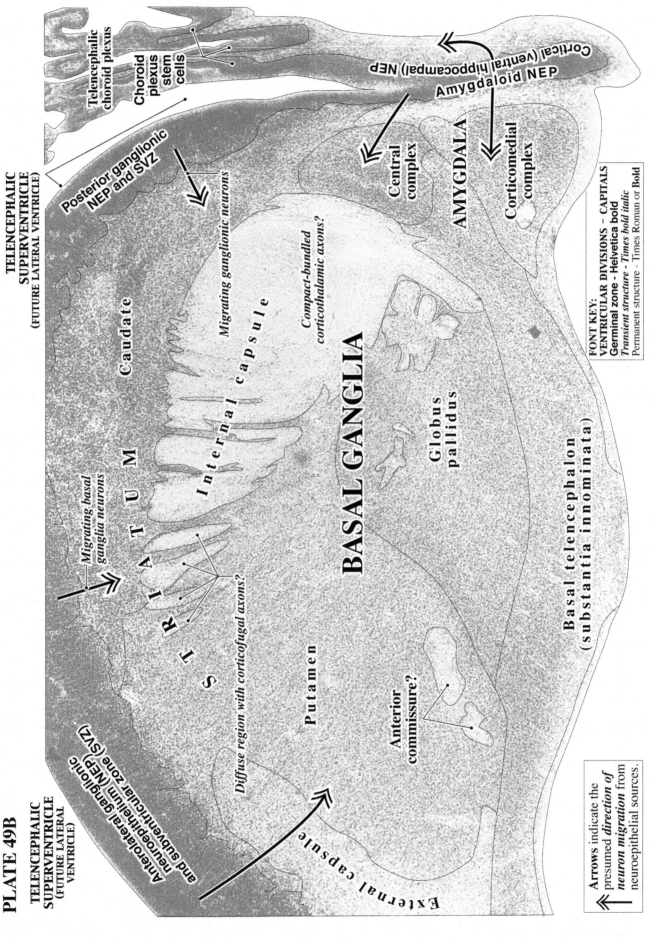

TELENCEPHALIC
SUPERVENTRICLE
(FUTURE LATERAL
VENTRICLE)

Anterolateral ganglionic
neuroepithelium (NEP) SVZ)
and subventricular zone

S T R I A T U M

Migrating basal
ganglia neurons

Diffuse region with corticofugal axons?

Putamen

External capsule

Anterior
commissure?

Globus
pallidus

BASAL GANGLIA

Internal capsule

Compact-bundled
corticothalamic axons?

Migrating ganglionic neurons

Caudate

Posterior ganglionic
NEP and SVZ

TELENCEPHALIC
SUPERVENTRICLE
(FUTURE LATERAL VENTRICLE)

Telencephalic
choroid plexus

Choroid
plexus
stem
cells

Basal telencephalon
(substantia innominata)

AMYGDALA

Central
complex

Corticomedial
complex

Amygdaloid NEP

Cortical (ventral hippocampal) NEP

FONT KEY:
VENTRICULAR DIVISIONS – CAPITALS
Germinal zone - Helvetica bold
Transient structure - Times bold italic
Permanent structure - Times Roman or Bold

Arrows indicate the presumed direction of neuron migration from neuroepithelial sources.

HYPOTHALAMUS AND PREOPTIC AREA

PLATE 50A

CR 33 mm, GW 9.6, C145
Sagittal, Slide 22, Section 2

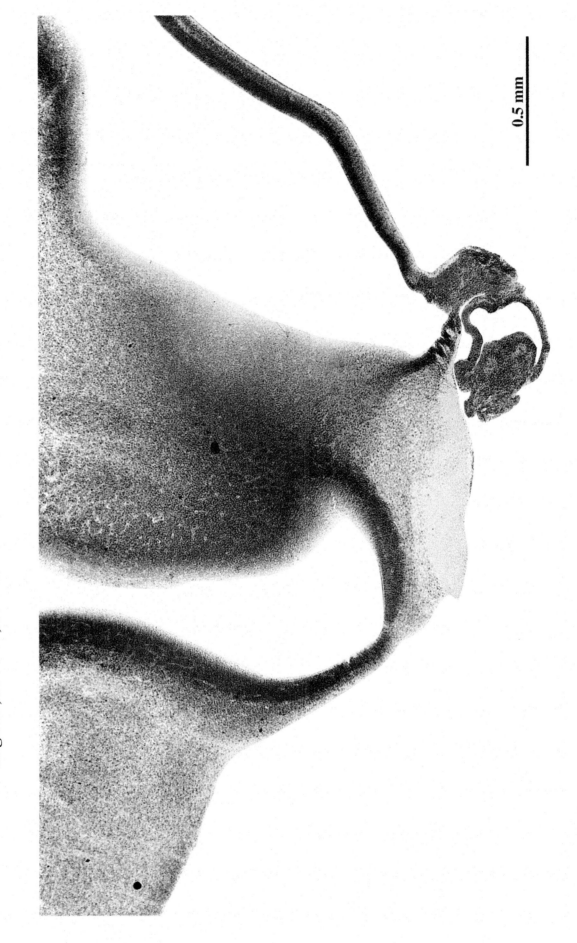

0.5 mm

See the entire section in Plates 40A-D.

PLATE 50B

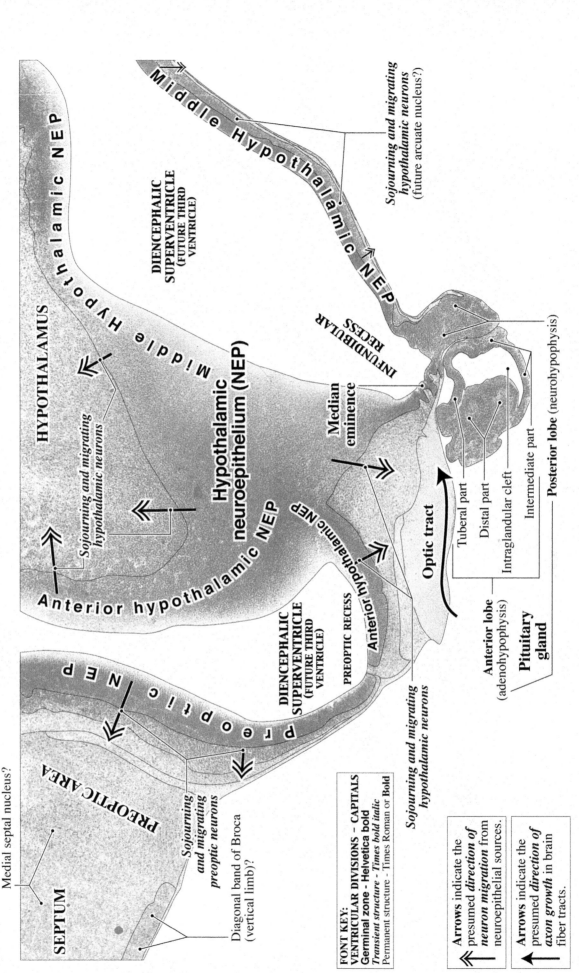

Middle Hypothalamic NEP

Sojourning and migrating hypothalamic neurons (future arcuate nucleus?)

DIENCEPHALIC SUPERVENTRICLE (FUTURE THIRD VENTRICLE)

INFUNDIBULAR RECESS

HYPOTHALAMUS

Middle Hypothalamic NEP

Sojourning and migrating hypothalamic neurons

Hypothalamic neuroepithelium (NEP)

Median eminence

Anterior hypothalamic NEP

Sojourning and migrating hypothalamic neurons

Anterior hypothalamic NEP

Optic tract

Tuberal part
Distal part
Intraglandular cleft
Intermediate part

Posterior lobe (neurohypophysis)

Anterior lobe (adenohypophysis)

Pituitary gland

DIENCEPHALIC SUPERVENTRICLE (FUTURE THIRD VENTRICLE)

PREOPTIC RECESS

Preoptic NEP

PREOPTIC AREA

Sojourning and migrating hypothalamic neurons

Sojourning and migrating preoptic neurons

Medial septal nucleus?

SEPTUM

Diagonal band of Broca (vertical limb)?

FONT KEY:
VENTRICULAR DIVISIONS – CAPITALS
Germinal zone - Helvetica bold
Transient structure - Times bold italic
Permanent structure - Times Roman or Bold

Arrows indicate the presumed *direction of neuron migration* from neuroepithelial sources.

Arrows indicate the presumed *direction of axon growth* in brain fiber tracts.

DIENCEPHALON AND BASAL TELENCEPHALON

PLATE 51A

CR 33 mm, GW 9.6,
C145, Sagittal,
Slide 20, Section 2

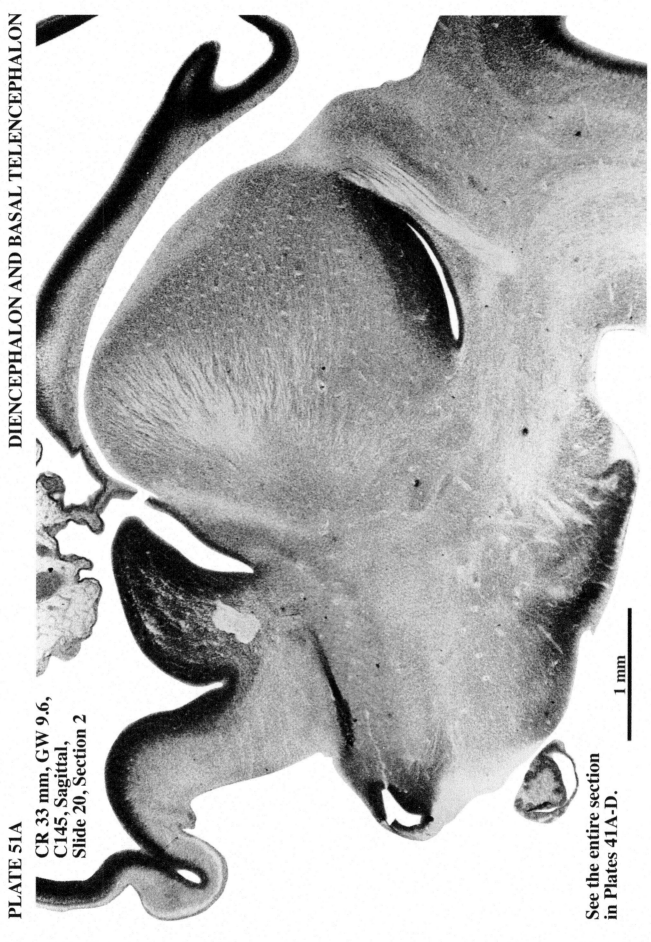

1 mm

See the entire section
in Plates 41A-D.

145

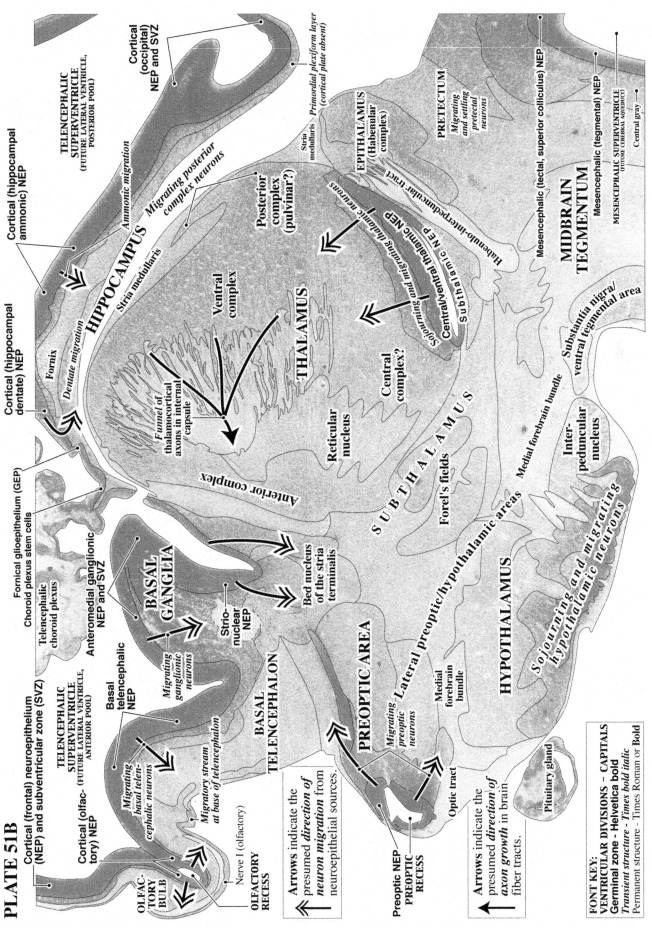

PLATE 51B

Cortical (frontal) neuroepithelium (NEP) and subventricular zone (SVZ)

TELENCEPHALIC SUPERVENTRICLE (FUTURE LATERAL VENTRICLE, ANTERIOR POOL)

Cortical (olfac-tory) NEP

Basal telencephalic NEP

Fornical glioepithelium (GEP)
Choroid plexus stem cells

Telencephalic choroid plexus

Anteromedial ganglionic NEP and SVZ

Migrating ganglionic neurons

Cortical (hippocampal ammonic) NEP

Cortical (hippocampal dentate) NEP

TELENCEPHALIC SUPERVENTRICLE (FUTURE LATERAL VENTRICLE, POSTERIOR POOL)

Cortical (occipital) NEP and SVZ

Ammonic migration

Migrating posterior complex neurons

HIPPOCAMPUS

Stria medullaris

Fornix

Dentate migration

Posterior complex (pulvinar?)

Ventral complex

Funnel of thalamocortical axons in internal capsule

Anterior complex

THALAMUS

Reticular nucleus

Central complex?

Stria medullaris – *Primordial plexiform layer (cortical plate absent)*

EPITHALAMUS ((Habenular complex)

PRETECTUM

Migrating and settling pretectal neurons

Habenulo-interpeduncular tract

Sojourning and migrating thalamic neurons

Central/ventral thalamic NEP

Subthalamic NEP

Mesencephalic (tectal, superior colliculus) NEP

MIDBRAIN TEGMENTUM

Mesencephalic (tegmental) NEP

MESENCEPHALIC SUPERVENTRICLE (FUTURE CEREBRAL AQUEDUCT)

Central gray

Substantia nigra/ventral tegmental area

S U B T H A L A M U S

Forel's fields

Medial forebrain bundle

Inter-peduncular nucleus

BASAL GANGLIA

Strio-nuclear NEP

Bed nucleus of the stria terminalis

Migrating basal telen-cephalic neurons

Migratory stream at base of telencephalon

BASAL TELENCEPHALON

Nerve I (olfactory)

OLFAC-TORY BULB

OLFACTORY RECESS

PREOPTIC AREA

Migrating preoptic neurons

Preoptic NEP
PREOPTIC RECESS

Lateral preoptic/hypothalamic areas

Medial forebrain bundle

HYPOTHALAMUS

Sojourning and migrating hypothalamic neurons

Optic tract

Pituitary gland

Arrows indicate the presumed *direction of neuron migration* from neuroepithelial sources.

Arrows indicate the presumed *direction of axon growth* in brain fiber tracts.

FONT KEY:
VENTRICULAR DIVISIONS – CAPITALS
Germinal zone - Helvetica bold
Transient structure - Times bold italic
Permanent structure - Times Roman or Bold

DIENCEPHALON AND BASAL TELENCEPHALON

PLATE 52A

CR 33 mm, GW 9.6,
C145, Sagittal,
Slide 19, Section 2

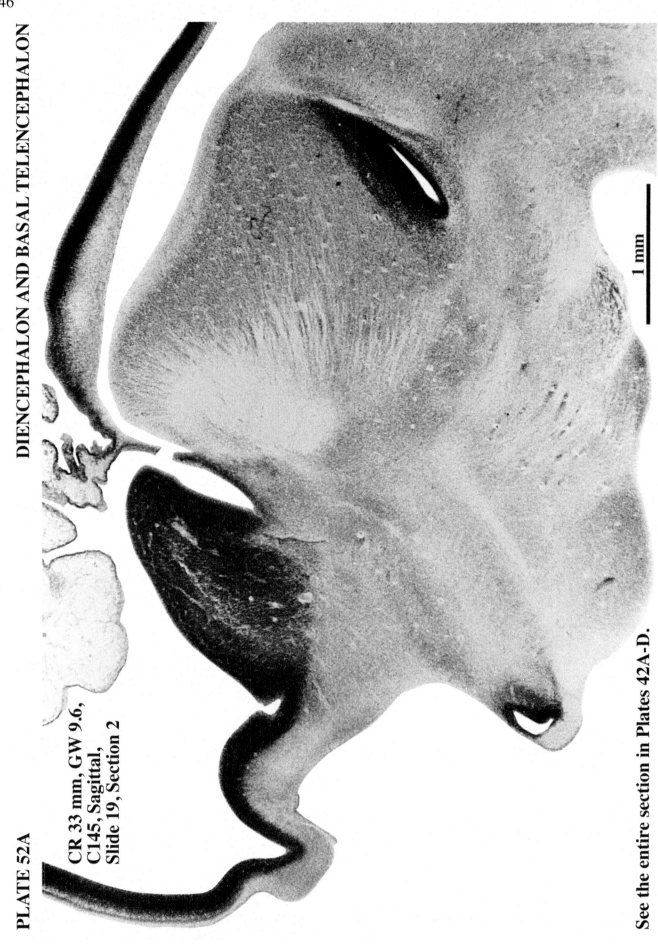

1 mm

See the entire section in Plates 42A-D.

PLATE 52B

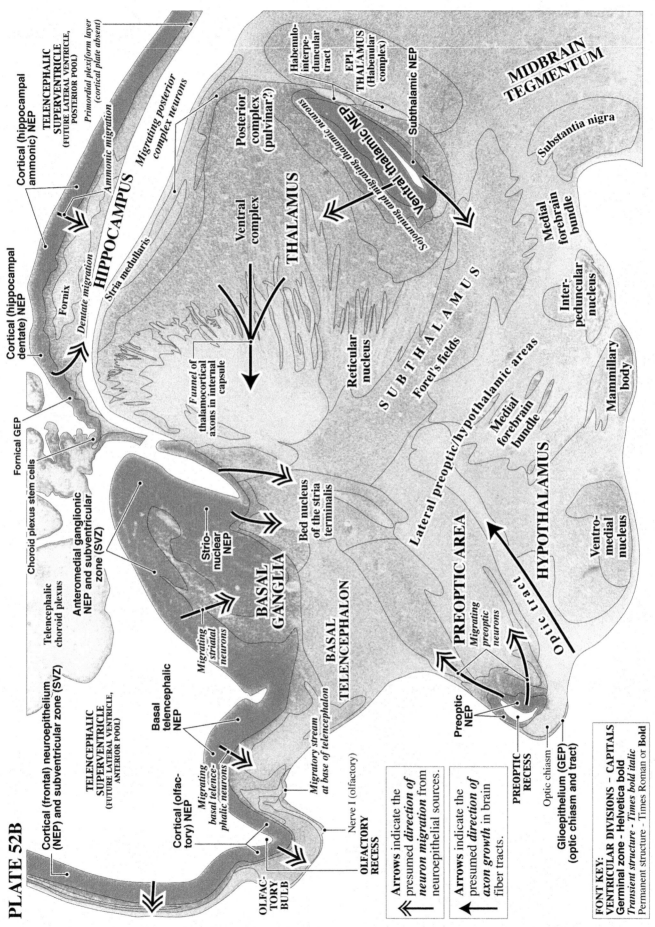

Cortical (hippocampal ammonic) NEP

TELENCEPHALIC SUPERVENTRICLE (FUTURE LATERAL VENTRICLE, POSTERIOR POOL)

Primordial plexiform layer (cortical plate absent)

Ammonic migration

Migrating posterior complex neurons

Cortical (hippocampal dentate) NEP

HIPPOCAMPUS

Stria medullaris

Fornix

Dentate migration

Fornical GEP

Choroid plexus stem cells

Telencephalic choroid plexus

Cortical (frontal) neuroepithelium (NEP) and subventricular zone (SVZ)

Anteromedial ganglionic NEP and subventricular zone (SVZ)

TELENCEPHALIC SUPERVENTRICLE (FUTURE LATERAL VENTRICLE, ANTERIOR POOL)

Basal telencephalic NEP

Migrating striatal neurons

Strio-nuclear NEP

BASAL GANGLIA

Bed nucleus of the stria terminalis

BASAL TELENCEPHALON

Migratory stream at base of telencephalon

Cortical (olfactory) NEP

Migrating basal telencephalic neurons

Nerve 1 (olfactory)

OLFACTORY BULB

OLFACTORY RECESS

Posterior complex (pulvinar?)

Ventral complex

THALAMUS

Sojourning and migrating thalamic neurons

Ventral thalamic NEP

Habenulo-interpeduncular tract

EPITHALAMUS (Habenular complex)

Subthalamic NEP

Funnel of thalamocortical axons in internal capsule

Reticular nucleus

SUBTHALAMUS

Forel's fields

Lateral preoptic/hypothalamic areas

PREOPTIC AREA

Migrating preoptic neurons

Preoptic NEP

PREOPTIC RECESS

Optic chiasm

Glioepithelium (GEP) (optic chiasm and tract)

Optic tract

Medial forebrain bundle

HYPOTHALAMUS

Ventromedial nucleus

MIDBRAIN TEGMENTUM

Substantia nigra

Medial forebrain bundle

Interpeduncular nucleus

Mammillary body

⇐ Arrows indicate the presumed *direction of neuron migration* from neuroepithelial sources.

← Arrows indicate the presumed *direction of axon growth* in brain fiber tracts.

FONT KEY:
VENTRICULAR DIVISIONS – CAPITALS
Germinal zone - Helvetica bold
Transient structure - Times bold italic
Permanent structure - Times Roman or Bold

DIENCEPHALON AND BASAL TELENCEPHALON

PLATE 53A   CR 33 mm, GW 9.6, C145,
Sagittal, Slide 18, Section 2

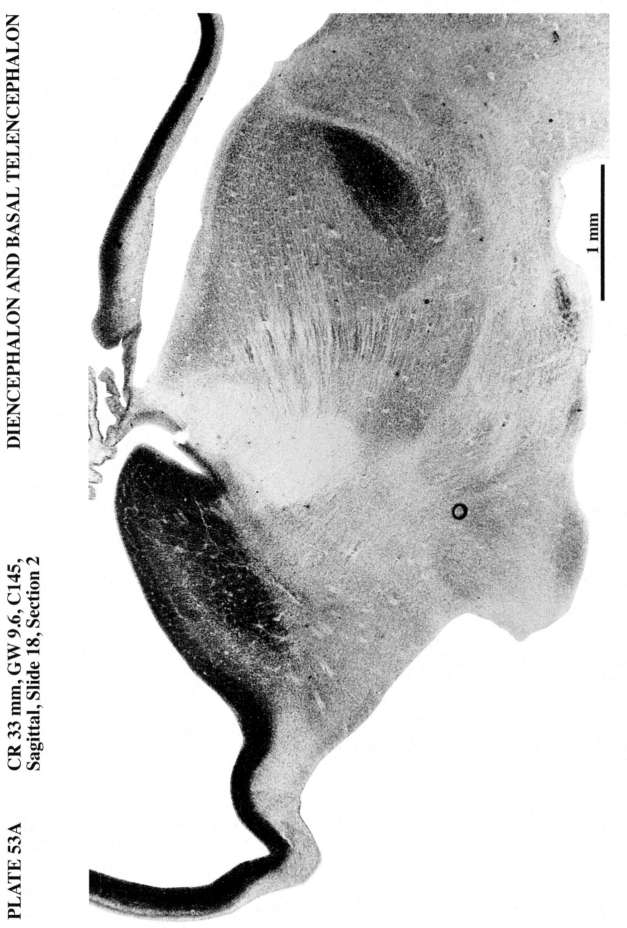

1 mm

See the entire section in Plates 43A-D.

PLATE 53B

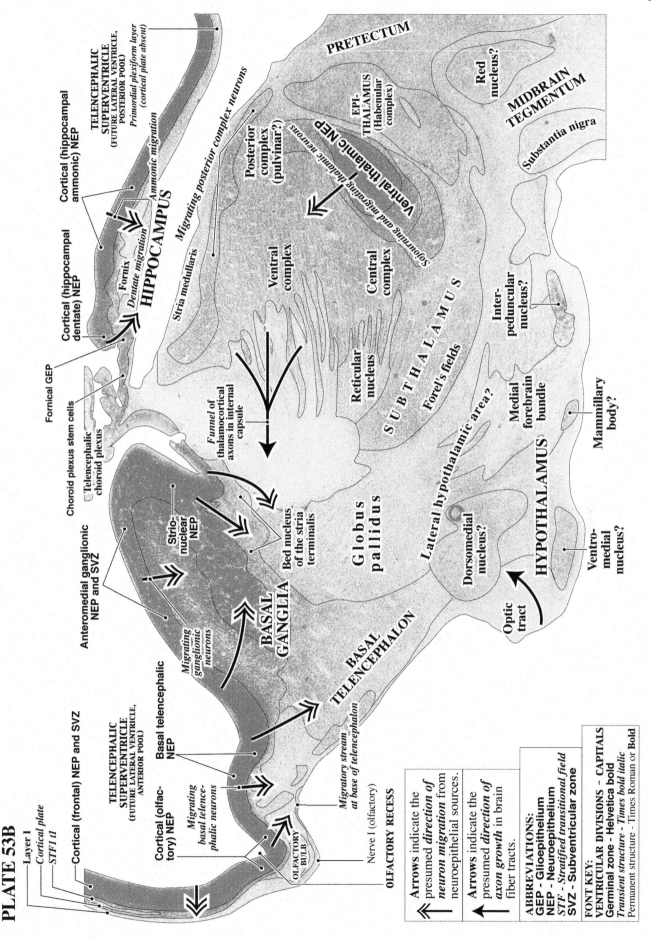

Layer 1
Cortical plate
STF1 t1
Cortical (frontal) NEP and SVZ

TELENCEPHALIC SUPERVENTRICLE
(FUTURE LATERAL VENTRICLE, ANTERIOR POOL)

Basal telencephalic NEP

Cortical (olfactory) NEP

Migrating basal telencephalic neurons

OLFACTORY BULB

Migratory stream at base of telencephalon

Nerve I (olfactory)

OLFACTORY RECESS

Anteromedial ganglionic NEP and SVZ

Strio-nuclear NEP

Migrating ganglionic neurons

BASAL GANGLIA

BASAL TELENCEPHALON

Bed nucleus of the stria terminalis

Funnel of thalamocortical axons in internal capsule

Telencephalic choroid plexus

Choroid plexus stem cells

Fornical GEP

Cortical (hippocampal dentate) NEP

Cortical (hippocampal ammonic) NEP

TELENCEPHALIC SUPERVENTRICLE
(FUTURE LATERAL VENTRICLE, POSTERIOR POOL)

Primordial plexiform layer (cortical plate absent)

Ammonic migration

Dentate migration

Fornix

Cortical (hippocampal dentate) NEP

HIPPOCAMPUS

Stria medullaris

Migrating posterior complex neurons

Posterior complex (pulvinar?)

PRETECTUM

EPI-THALAMUS (Habenular complex)

Ventral complex

Ventral thalamic NEP

Sojourning and migrating thalamic neurons

Central complex

Reticular nucleus

S U B T H A L A M U S

Forel's fields

G l o b u s p a l l i d u s

Lateral hypothalamic area?

Dorsomedial nucleus?

Optic tract

HYPOTHALAMUS

Ventro-medial nucleus?

Mammillary body?

Medial forebrain bundle

Inter-peduncular nucleus?

Substantia nigra

MIDBRAIN TEGMENTUM

Red nucleus?

Arrows indicate the presumed *direction of neuron migration* from neuroepithelial sources.

Arrows indicate the presumed *direction of axon growth* in brain fiber tracts.

ABBREVIATIONS:
GEP - Glioepithelium
NEP - Neuroepithelium
STF - Stratified transitional field
SVZ - Subventricular zone

FONT KEY:
VENTRICULAR DIVISIONS – CAPITALS
Germinal zone - Helvetica bold
Transient structure - Times bold italic
Permanent structure - Times Roman or **Bold**

DIENCEPHALON AND BASAL TELENCEPHALON

PLATE 54A

CR 33 mm, GW 9.6, C145,
Sagittal, Slide 17, Section 2

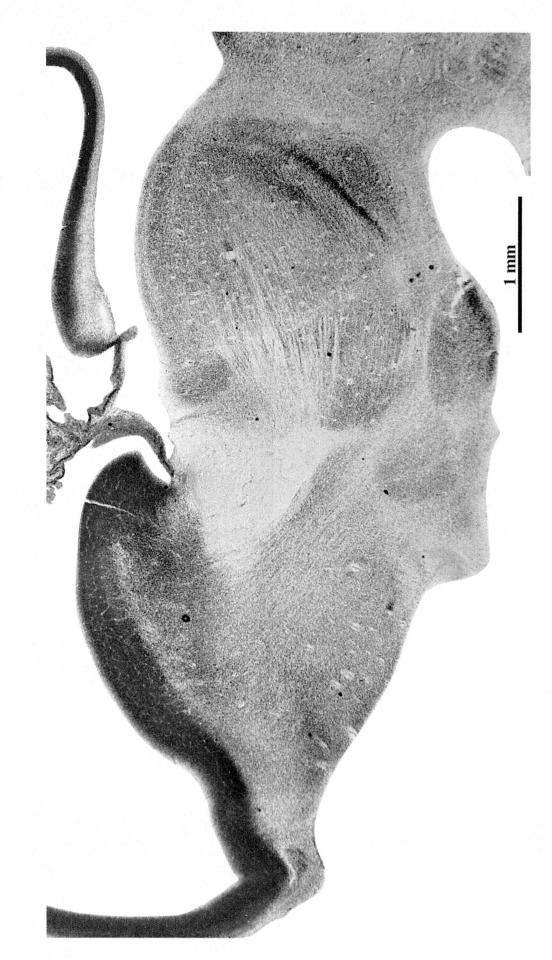

1 mm

See nearby complete sections in Plates 43A-D and 44A-D.

PLATE 54B

151

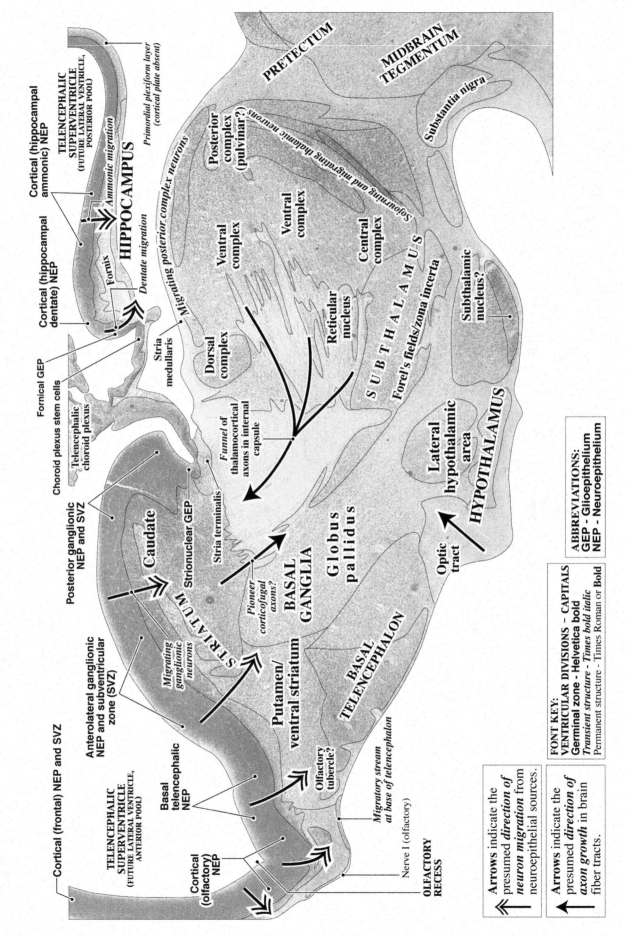

Arrows indicate the *presumed direction of neuron migration* from neuroepithelial sources.

Arrows indicate the *presumed direction of axon growth* in brain fiber tracts.

FONT KEY:
VENTRICULAR DIVISIONS – CAPITALS
Germinal zone - Helvetica bold
Transient structure - Times bold italic
Permanent structure - Times Roman or Bold

ABBREVIATIONS:
GEP - Glioepithelium
NEP - Neuroepithelium

152

MIDBRAIN TECTUM AND TEGMENTUM

PLATE 55A

CR 33 mm, GW 9.6, C145,
Sagittal, Slide 24, Section 2

See the entire section
in Plates 39A-D.

1 mm

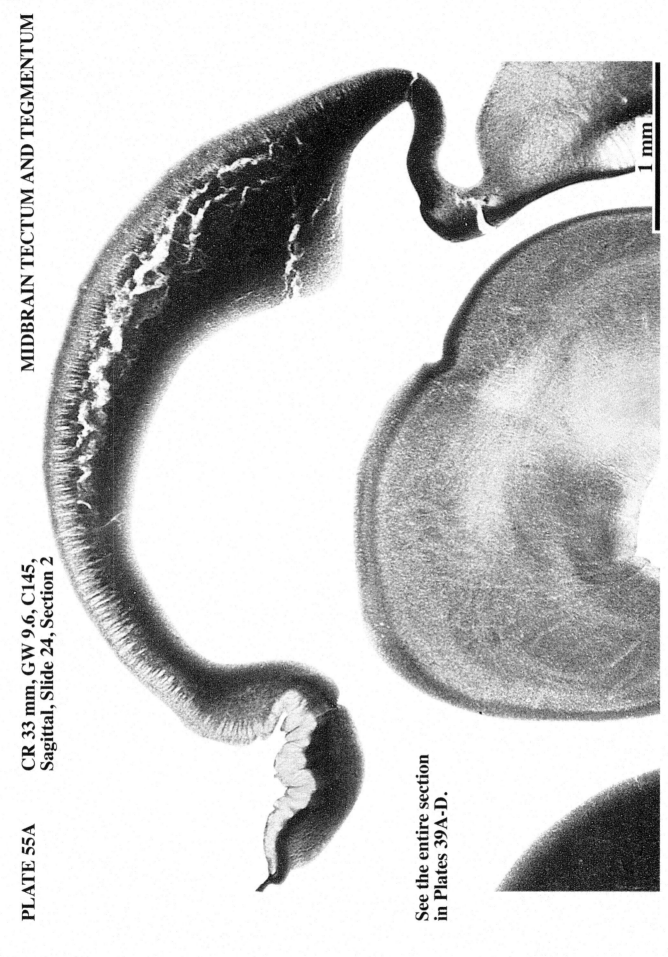

153

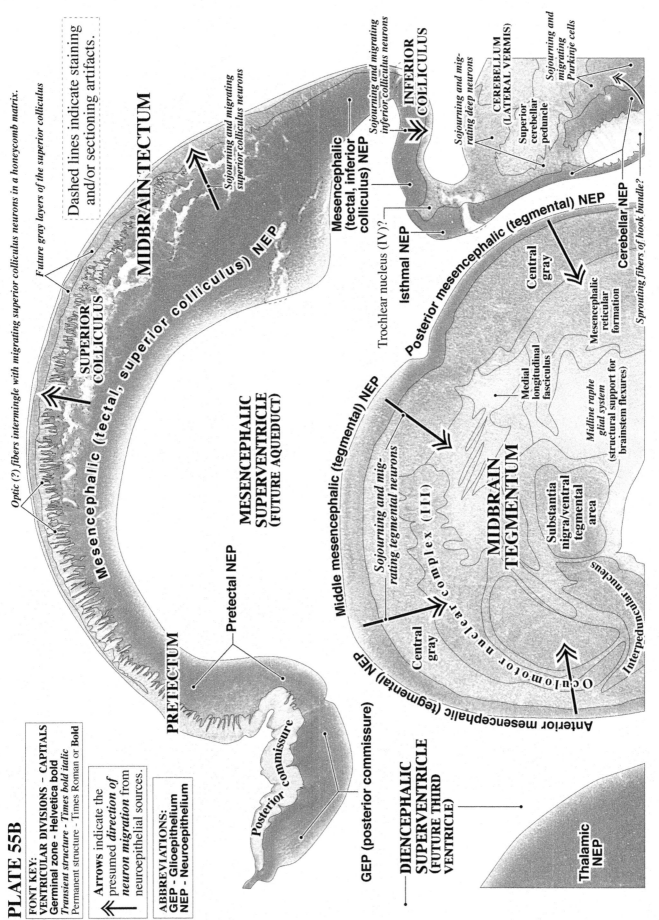

PLATE 55B

FONT KEY:
VENTRICULAR DIVISIONS - CAPITALS
Germinal zone - Helvetica bold
Transient structure - Times bold italic
Permanent structure - Times Roman or **Bold**

⟸ Arrows indicate the presumed *direction of neuron migration* from neuroepithelial sources.

ABBREVIATIONS:
GEP - Glioepithelium
NEP - Neuroepithelium

Dashed lines indicate staining and/or sectioning artifacts.

Optic (?) fibers intermingle with migrating superior colliculus neurons in a honeycomb matrix.

Future gray layers of the superior colliculus

MIDBRAIN TECTUM

Sojourning and migrating superior colliculus neurons

SUPERIOR COLLICULUS

Mesencephalic (tectal, superior colliculus) NEP

PRETECTUM

Pretectal NEP

MESENCEPHALIC SUPERVENTRICLE (FUTURE AQUEDUCT)

Posterior commissure

GEP (posterior commissure)

DIENCEPHALIC SUPERVENTRICLE (FUTURE THIRD VENTRICLE)

Thalamic NEP

Mesencephalic (tectal, inferior colliculus) NEP

Sojourning and migrating inferior colliculus neurons

INFERIOR COLLICULUS

Sojourning and migrating deep neurons

CEREBELLUM (LATERAL VERMIS)

Superior cerebellar peduncle

Sojourning and migrating Purkinje cells

Cerebellar NEP

Sprouting fibers of hook bundle?

Trochlear nucleus (IV)?

Isthmal NEP

Posterior mesencephalic (tegmental) NEP

Central gray

Mesencephalic reticular formation

Medial longitudinal fasciculus

Midline raphe glial system (structural support for brainstem flexures)

MIDBRAIN TEGMENTUM

Substantia nigra/ventral tegmental area

Interpeduncular nucleus

Middle mesencephalic (tegmental) NEP

Sojourning and migrating tegmental neurons

Central gray

Oculomotor complex (III)

Anterior mesencephalic (tegmental) NEP

MIDBRAIN TEGMENTUM

PLATE 56A   CR 33 mm, GW 9.6, C145,
Sagittal, Slide 23, Section 2

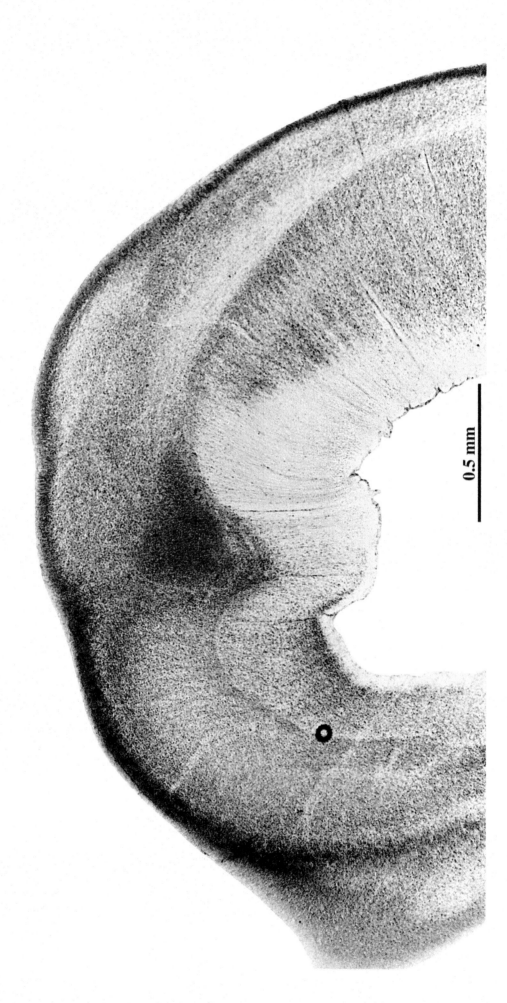

0.5 mm

See the entire section in Plates 39A-D.

PLATE 56B

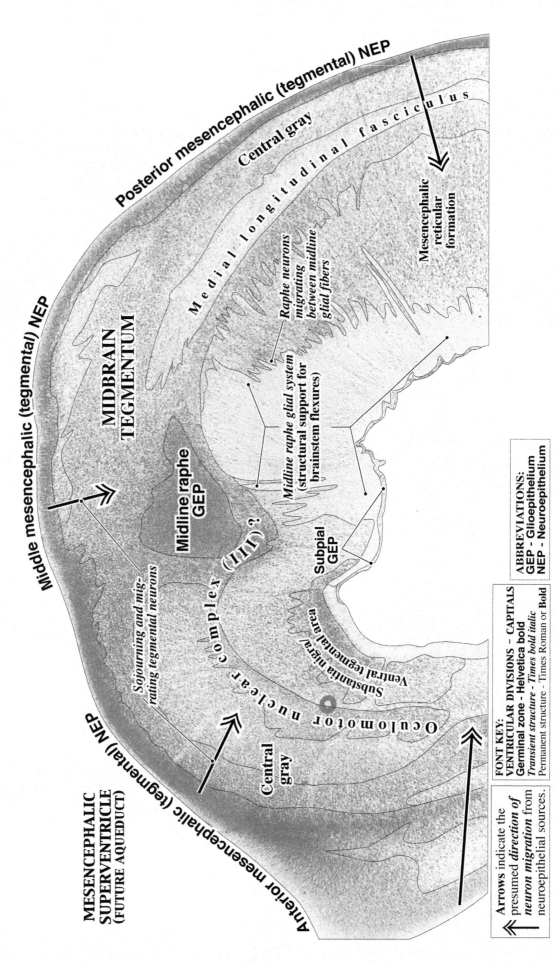

Posterior mesencephalic (tegmental) NEP

Central gray

Medial longitudinal fasciculus

Mesencephalic reticular formation

Raphe neurons migrating between midline glial fibers

Middle mesencephalic (tegmental) NEP

MIDBRAIN TEGMENTUM

Midline raphe glial system (structural support for brainstem flexures)

Midline raphe GEP

MESENCEPHALIC SUPERVENTRICLE (FUTURE AQUEDUCT)

Oculomotor nuclear complex (III)?

Sojourning and mig- rating tegmental neurons

Subpial GEP

Central gray

Substantia nigra/ Ventral tegmental area

Anterior mesencephalic (tegmental) NEP

 Arrows indicate the presumed *direction of neuron migration* from neuroepithelial sources.

FONT KEY:
VENTRICULAR DIVISIONS – CAPITALS
Germinal zone – Helvetica bold
Transient structure - Times bold italic
Permanent structure - Times Roman or Bold

ABBREVIATIONS:
GEP - Glioepithelium
NEP - Neuroepithelium

CEREBELLUM: LATERAL VERMIS

PLATE 57A

CR 33 mm, GW 9.6,
C145, Sagittal,
Slide 20,
Section 2

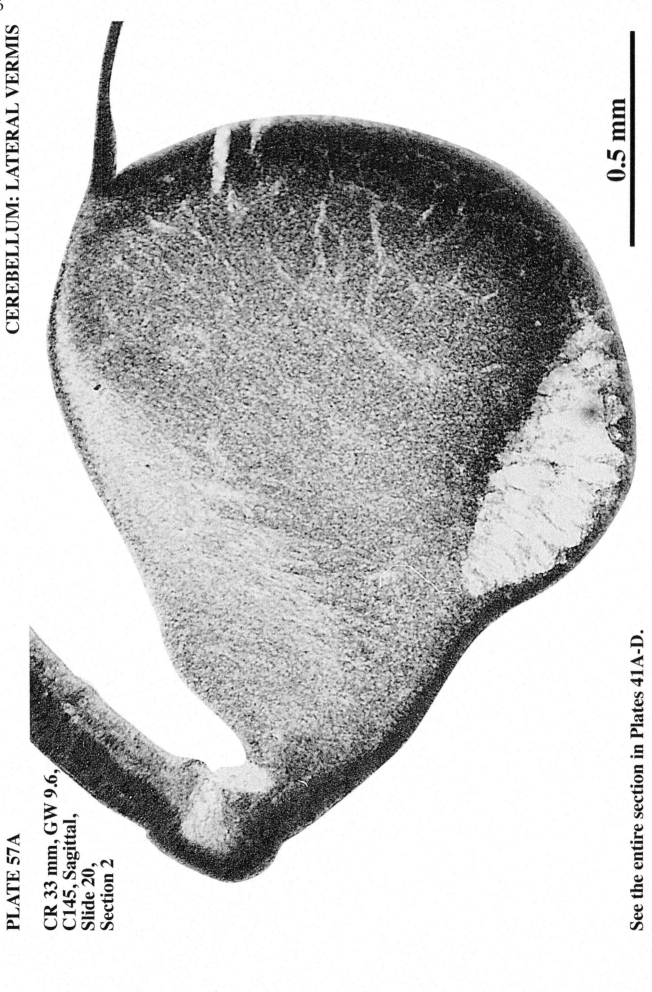

0.5 mm

See the entire section in Plates 41A-D.

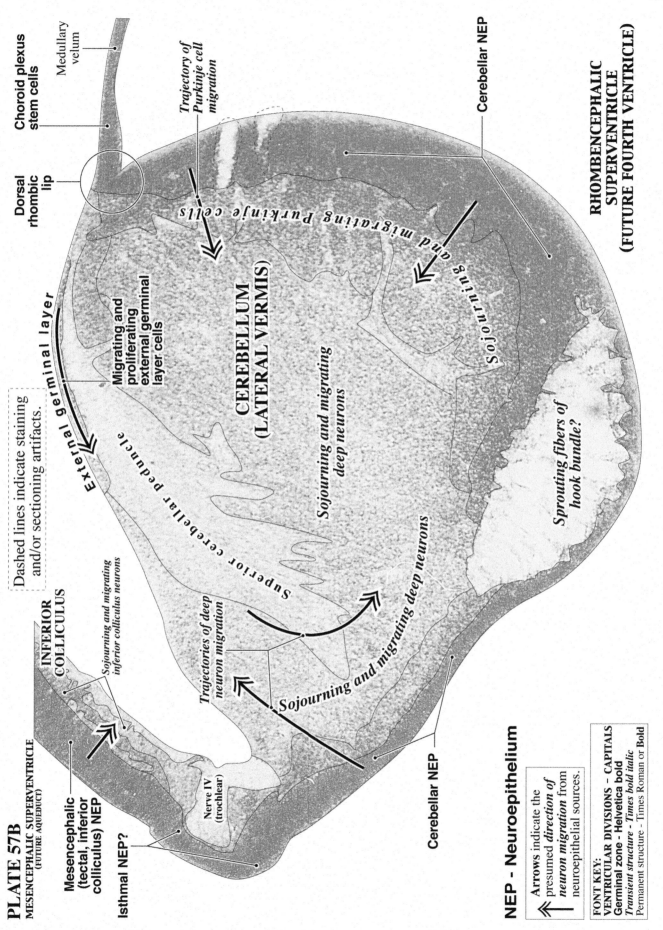

PLATE 57B
MESENCEPHALIC SUPERVENTRICLE
(FUTURE AQUEDUCT)

Choroid plexus stem cells

Medullary velum

Trajectory of Purkinje cell migration

Cerebellar NEP

Dorsal rhombic lip

RHOMBENCEPHALIC SUPERVENTRICLE
(FUTURE FOURTH VENTRICLE)

Sojourning and migrating Purkinje cells

Dashed lines indicate staining and/or sectioning artifacts.

External germinal layer

Migrating and proliferating external germinal layer cells

Superior cerebellar peduncle

CEREBELLUM
(LATERAL VERMIS)

Sojourning and migrating deep neurons

Sprouting fibers of hook bundle?

INFERIOR COLLICULUS

Sojourning and migrating inferior colliculus neurons

Trajectories of deep neuron migration

Sojourning and migrating deep neurons

Mesencephalic (tectal, inferior colliculus) NEP

Isthmal NEP?

Nerve IV (trochlear)

Cerebellar NEP

Cerebellar NEP

NEP - Neuroepithelium

Arrows indicate the presumed *direction of neuron migration* from neuroepithelial sources.

FONT KEY:
VENTRICULAR DIVISIONS – CAPITALS
Germinal zone - **Helvetica bold**
Transient structure - Times bold italic
Permanent structure - Times Roman or **Bold**

158

CEREBELLUM: HEMISPHERE

PLATE 58A

CR 33 mm, GW 9.6, C145,
Sagittal, Slide 12, Section 4

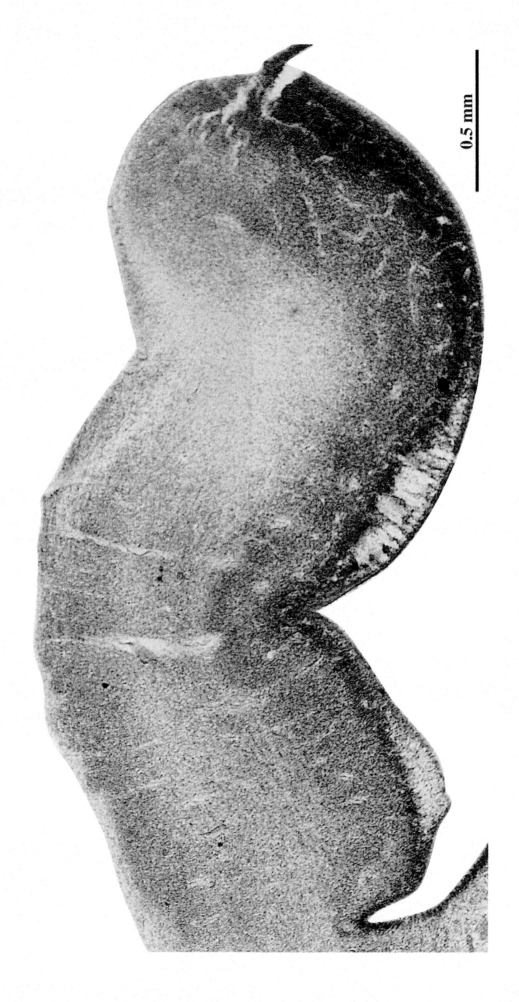

See the entire section in Plates 47A-D.

159

PLATE 58B

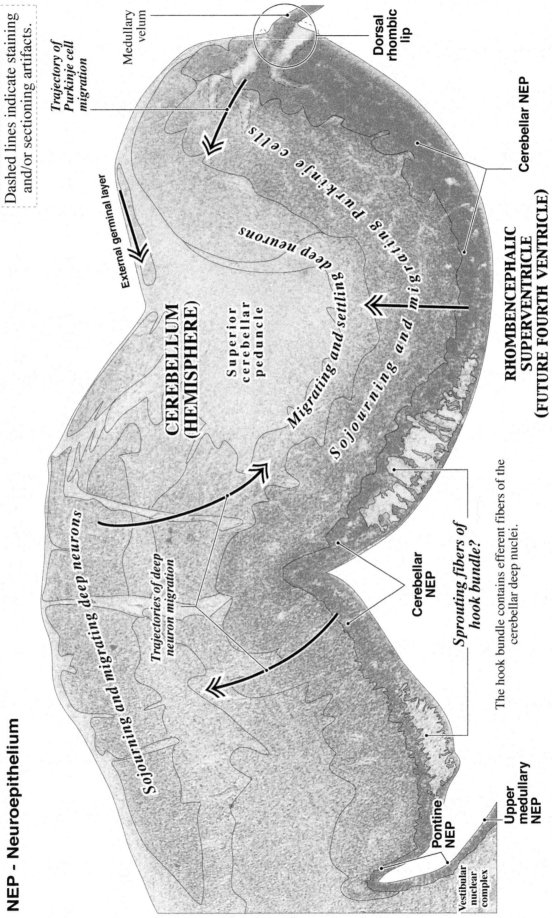

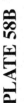

Arrows indicate the presumed *direction of neuron migration* from neuroepithelial sources.

FONT KEY:
VENTRICULAR DIVISIONS – CAPITALS
Germinal zone - **Helvetica bold**
Transient structure - *Times bold italic*
Permanent structure - Times Roman or **Bold**

NEP - Neuroepithelium

Dashed lines indicate staining and/or sectioning artifacts.

Trajectory of Purkinje cell migration

Medullary velum

Dorsal rhombic lip

Cerebellar NEP

External germinal layer

Migrating Purkinje cells

Sojourning and migrating deep neurons

Migrating and settling deep neurons

CEREBELLUM (HEMISPHERE)

Superior cerebellar peduncle

Sojourning and migrating deep neurons

Trajectories of deep neuron migration

RHOMBENCEPHALIC SUPERVENTRICLE (FUTURE FOURTH VENTRICLE)

Cerebellar NEP

Sprouting fibers of hook bundle?

The hook bundle contains efferent fibers of the cerebellar deep nuclei.

Pontine NEP

Upper medullary NEP

Vestibular nuclear complex

PONS AND MEDULLA

CR 33 mm, GW 9.6, C145,
Sagittal, Slide 23, Section 2

PLATE 59A

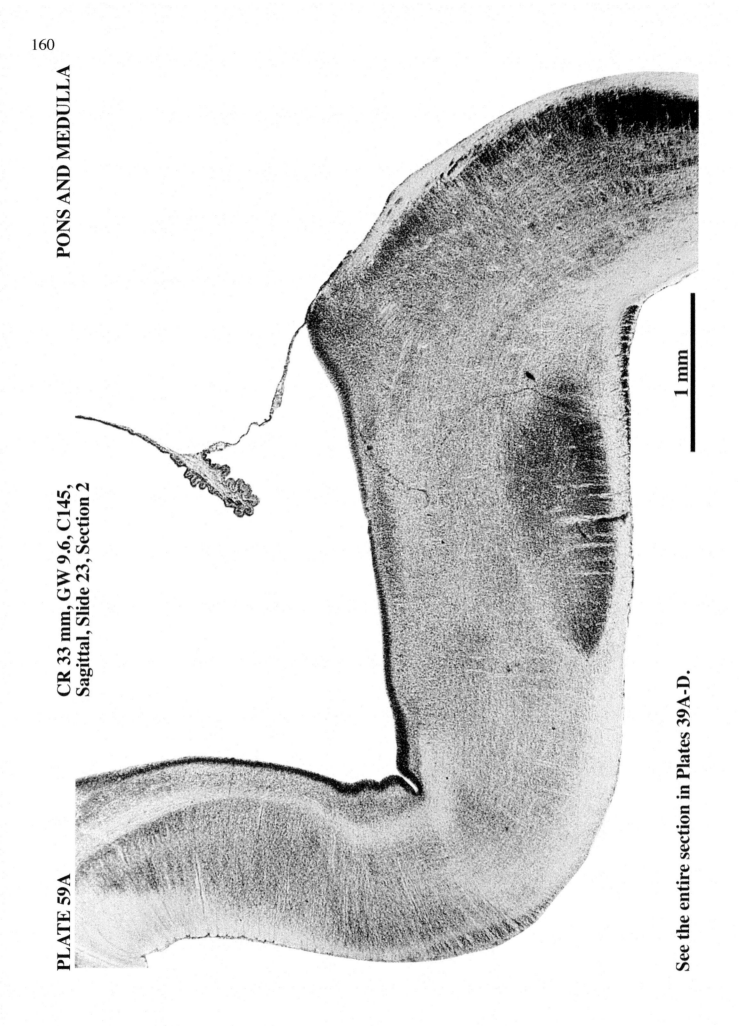

See the entire section in Plates 39A-D.

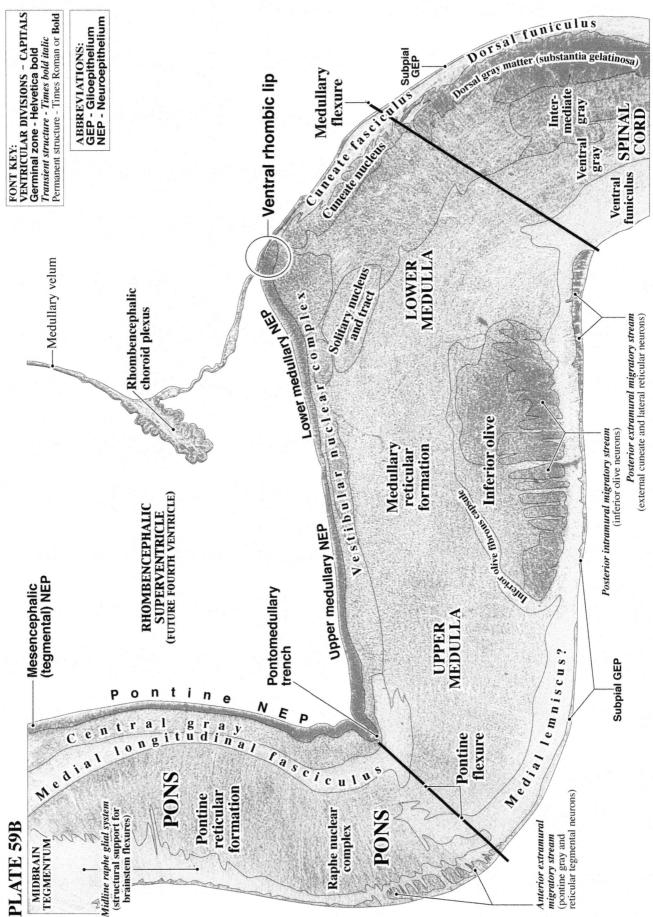

PLATE 59B

FONT KEY:
VENTRICULAR DIVISIONS – CAPITALS
Germinal zone - Helvetica bold
Transient structure - Times bold italic
Permanent structure - Times Roman or **Bold**

ABBREVIATIONS:
GEP - Glioepithelium
NEP - Neuroepithelium

Medullary velum

Rhombencephalic choroid plexus

Mesencephalic (tegmental) NEP

MIDBRAIN TEGMENTUM

Midline raphe glial system (structural support for brainstem flexures)

RHOMBENCEPHALIC SUPERVENTRICLE
(FUTURE FOURTH VENTRICLE)

Pontine NEP

Central gray

Medial longitudinal fasciculus

PONS

Pontine reticular formation

Raphe nuclear complex

PONS

Pontomedullary trench

Upper medullary NEP

Vestibular nuclear complex

Lower medullary NEP

Ventral rhombic lip

Medullary flexure

Cuneate fasciculus

Cuneate nucleus

Solitary nucleus and tract

Medullary reticular formation

LOWER MEDULLA

UPPER MEDULLA

Pontine flexure

Inferior olive

Inferior olive

Inferior olive fibrous capsule

Medial lemniscus?

Anterior extramural migratory stream (pontine gray and reticular tegmental neurons)

Subpial GEP

Posterior intramural migratory stream (inferior olive neurons)

Posterior extramural migratory stream (external cuneate and lateral reticular neurons)

Dorsal funiculus

Subpial GEP

Dorsal gray matter (substantia gelatinosa)

Inter-mediate

Ventral gray

SPINAL CORD

Ventral funiculus

PONS AND MEDULLA

PLATE 60A

CR 33 mm, GW 9.6, C145,
Sagittal, Slide 22, Section 2

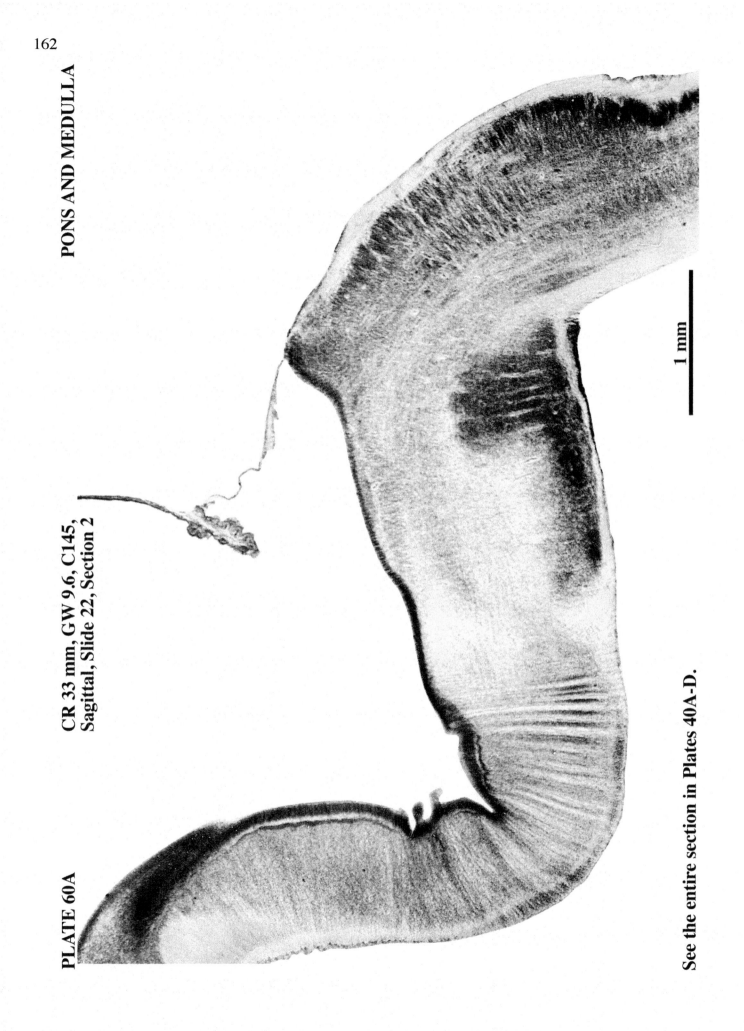

1 mm

See the entire section in Plates 40A-D.

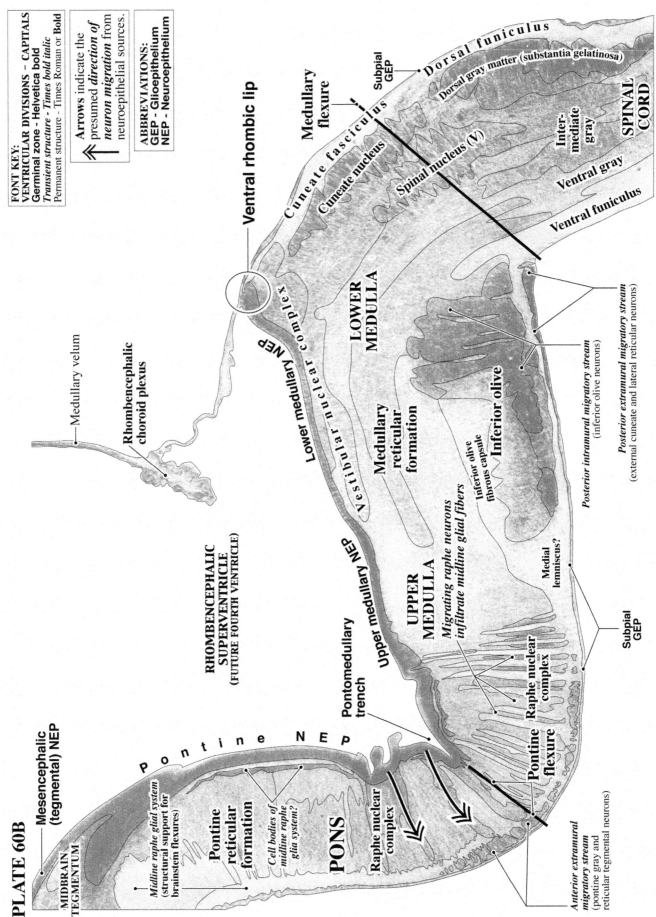

163

PLATE 60B

Mesencephalic (tegmental) NEP

MIDBRAIN TEGMENTUM

Pontine NEP

Midline raphe glial system (structural support for brainstem flexures)

Pontine reticular formation

Cell bodies of midline raphe glia system?

PONS

Raphe nuclear complex

Anterior extramural migratory stream (pontine gray and reticular tegmental neurons)

Medullary velum

Rhombencephalic choroid plexus

RHOMBENCEPHALIC SUPERVENTRICLE (FUTURE FOURTH VENTRICLE)

Pontomedullary trench

Upper medullary NEP

UPPER MEDULLA

Migrating raphe neurons infiltrate midline glial fibers

Raphe nuclear complex

Pontine flexure

Medial lemniscus?

Subpial GEP

Ventral rhombic lip

NEP

Lower medullary nuclear complex

Vestibular nuclear complex

LOWER MEDULLA

Medullary reticular formation

Inferior olive

Inferior olive capsule fibrous fibers

Posterior extramural migratory stream (external cuneate and lateral reticular neurons)

Posterior intramural migratory stream (inferior olive neurons)

Medullary flexure

Cuneate fasciculus

Cuneate nucleus

Spinal nucleus (V)

Subpial GEP

Dorsal funiculus

Dorsal gray matter (substantia gelatinosa)

Inter-mediate gray

SPINAL CORD

Ventral gray

Ventral funiculus

BRAINSTEM

PLATE 61A

CR 33 mm, GW 9.6, C145,
Sagittal, Slide 21, Section 2

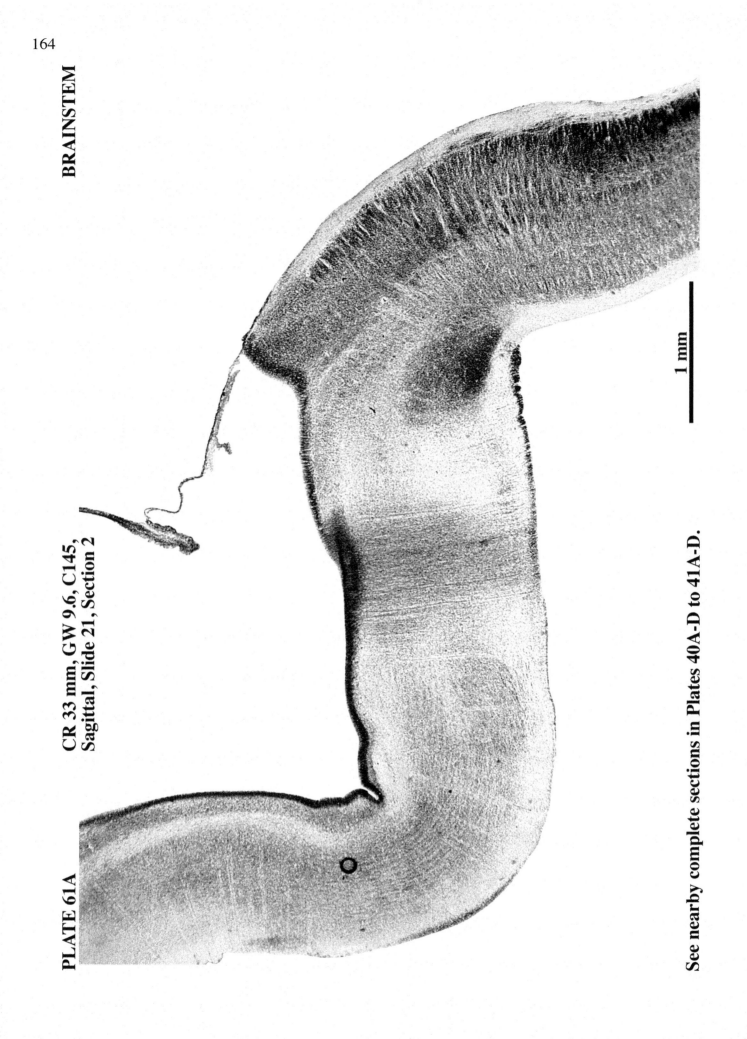

1 mm

See nearby complete sections in Plates 40A-D to 41A-D.

165

PLATE 61B

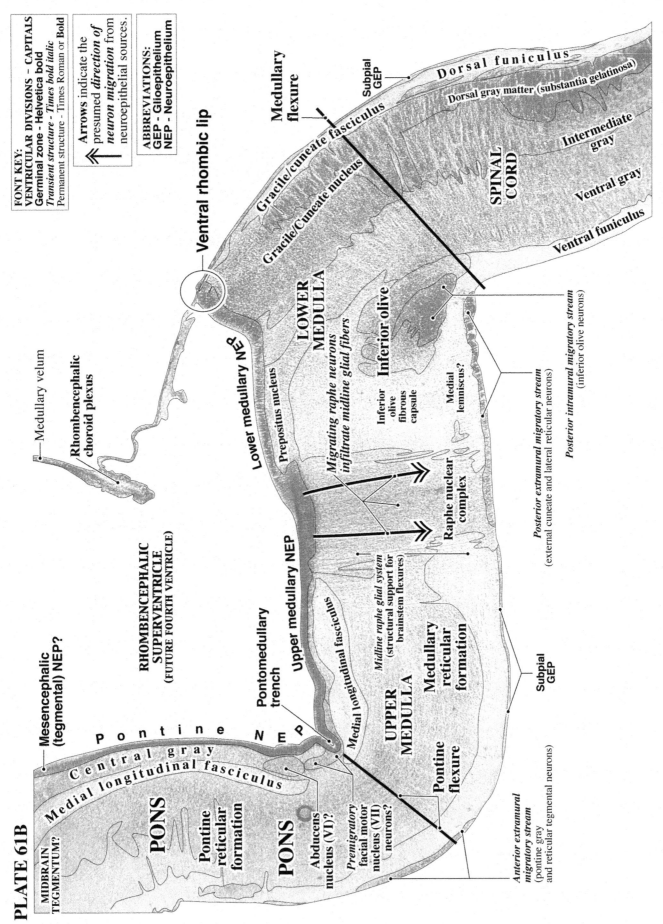

FONT KEY:
VENTRICULAR DIVISIONS – CAPITALS
Germinal zone – Helvetica bold
Transient structure – Times bold italic
Permanent structure – Times Roman or **Bold**

← Arrows indicate the presumed *direction of neuron migration* from neuroepithelial sources.

ABBREVIATIONS:
GEP - Glioepithelium
NEP - Neuroepithelium

Medullary velum

Rhombencephalic choroid plexus

Medullary flexure

Ventral rhombic lip

Gracile/cuneate fasciculus

Gracile/Cuneate nucleus

Subpial GEP

Dorsal funiculus

Dorsal gray matter (substantia gelatinosa)

Intermediate gray

SPINAL CORD

Ventral gray

Ventral funiculus

LOWER MEDULLA

Inferior olive

Lower medullary NEP

Prepositus nucleus

Migrating raphe neurons infiltrate midline glial fibers

Inferior olive fibrous capsule

Medial lemniscus?

Raphe nuclear complex

Posterior extramural migratory stream
(external cuneate and lateral reticular neurons)

Posterior intramural migratory stream
(inferior olive neurons)

MIDBRAIN TEGMENTUM?

Mesencephalic (tegmental) NEP?

RHOMBENCEPHALIC SUPERVENTRICLE
(FUTURE FOURTH VENTRICLE)

Upper medullary NEP

Pontomedullary trench

Medial longitudinal fasciculus

Midline raphe glial system
(structural support for brainstem flexures)

UPPER MEDULLA

Medullary reticular formation

Subpial GEP

Pontine flexure

P o n t i n e   N E P

Central gray

Medial longitudinal fasciculus

PONS

Pontine reticular formation

PONS

Abducens nucleus (VI)?

Premigratory facial motor nucleus (VII) neurons?

Anterior extramural migratory stream
(pontine gray and reticular tegmental neurons)

BRAINSTEM

PLATE 62A

CR 33 mm, GW 9.6, C145,
Sagittal, Slide 20, Section 2

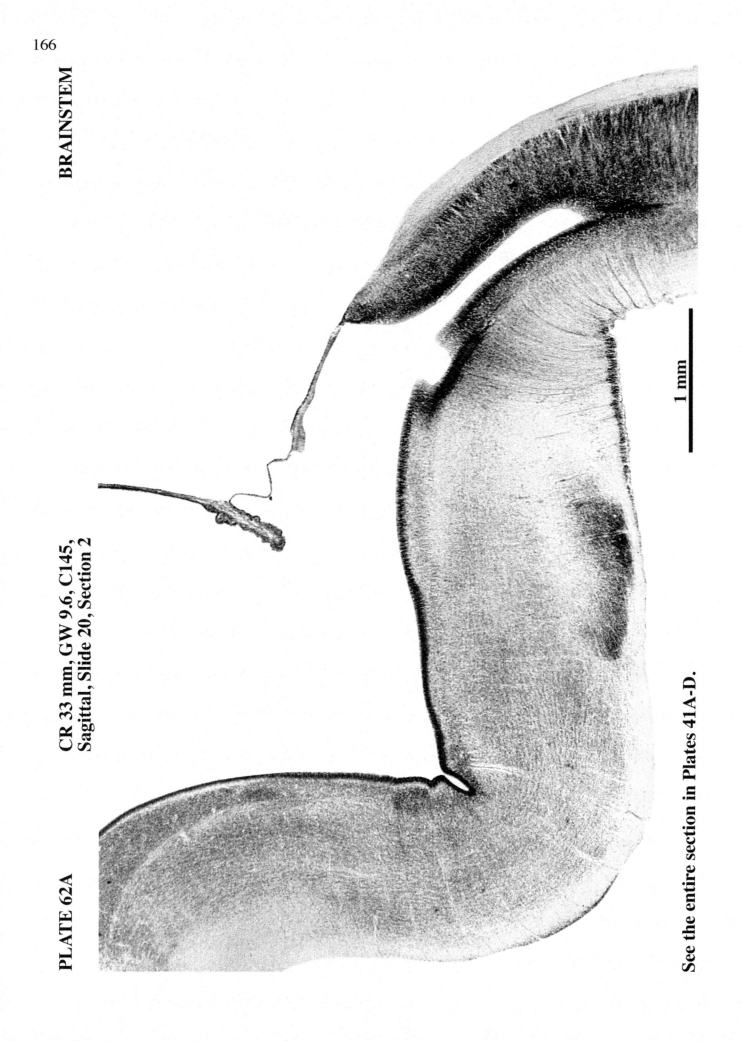

1 mm

See the entire section in Plates 41A-D.

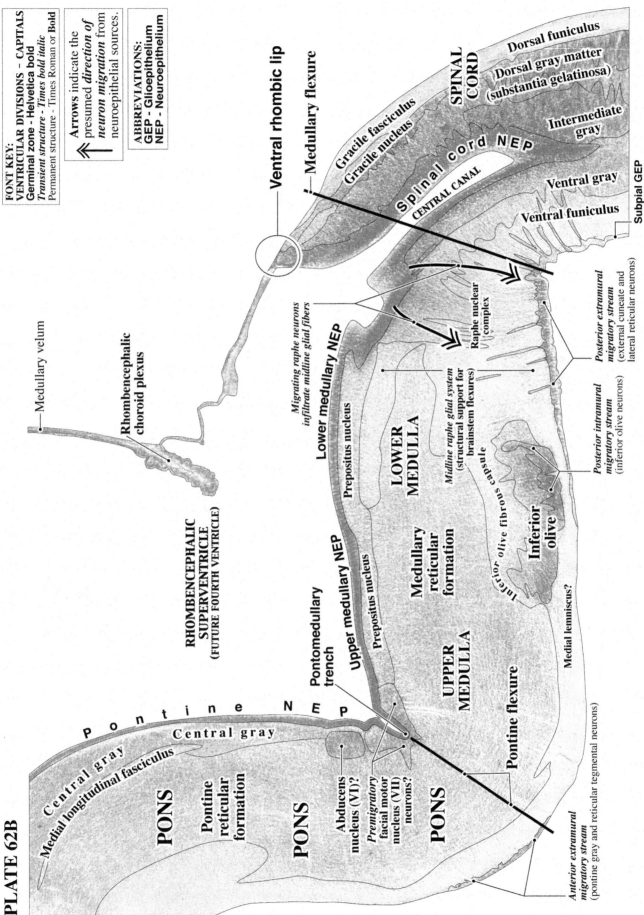

PLATE 62B

FONT KEY:
VENTRICULAR DIVISIONS - CAPITALS
Germinal zone - Helvetica bold
Transient structure - Times bold italic
Permanent structure - Times Roman or **Bold**

⟵ **Arrows** indicate the presumed *direction of neuron migration* from neuroepithelial sources.

ABBREVIATIONS:
GEP - Glioepithelium
NEP - Neuroepithelium

Medullary velum

Rhombencephalic choroid plexus

RHOMBENCEPHALIC SUPERVENTRICLE (FUTURE FOURTH VENTRICLE)

P o n t i n e   N E P

Central gray

C e n t r a l   g r a y

Medial longitudinal fasciculus

PONS

Pontine reticular formation

PONS

Abducens nucleus (VI)?

Premigratory facial motor nucleus (VII) neurons?

PONS

Pontomedullary trench

Upper medullary NEP

Prepositus nucleus

Anterior extramural migratory stream (pontine gray and reticular tegmental neurons)

Pontine flexure

UPPER MEDULLA

Medial lemniscus?

Inferior olive fibrous capsule

Inferior olive

Medullary reticular formation

LOWER MEDULLA

Midline raphe glial system (structural support for brainstem flexures)

Prepositus nucleus

Lower medullary NEP

Migrating raphe neurons infiltrate midline glial fibers

Raphe nuclear complex

Posterior intramural migratory stream (inferior olive neurons)

Posterior extramural migratory stream (external cuneate and lateral reticular neurons)

Subpial GEP

Ventral funiculus

Ventral gray

Intermediate gray

CENTRAL CANAL

Spinal cord NEP

Gracile nucleus

Gracile fasciculus

Ventral rhombic lip

Medullary flexure

Dorsal gray matter (substantia gelatinosa)

Dorsal funiculus

SPINAL CORD

LATERAL PONS, MEDULLA, AND PERIPHERAL GANGLIA

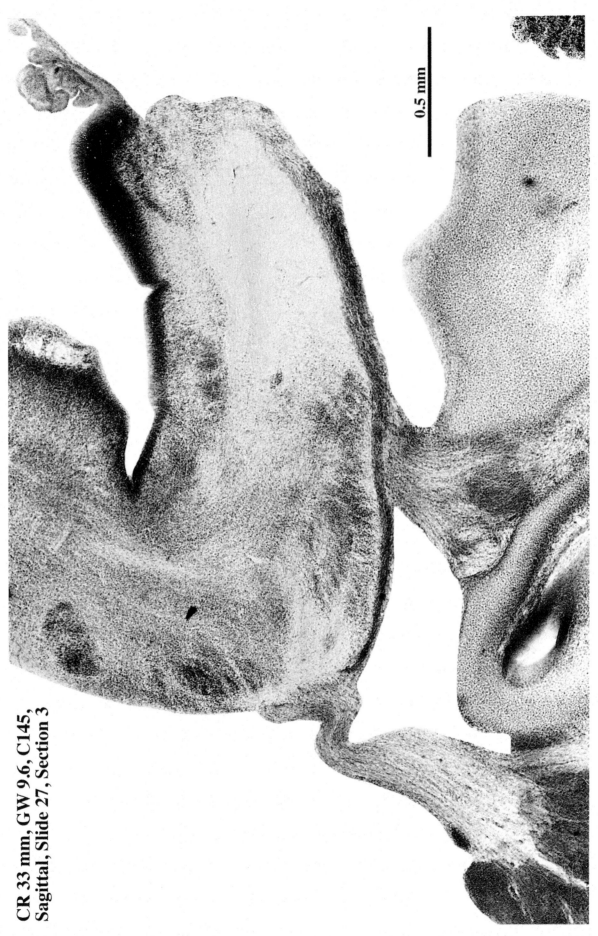

PLATE 63A

CR 33 mm, GW 9.6, C145,
Sagittal, Slide 27, Section 3

0.5 mm

See similar areas from the left side of the brain in Plates 46A-D to 48A-D.

169

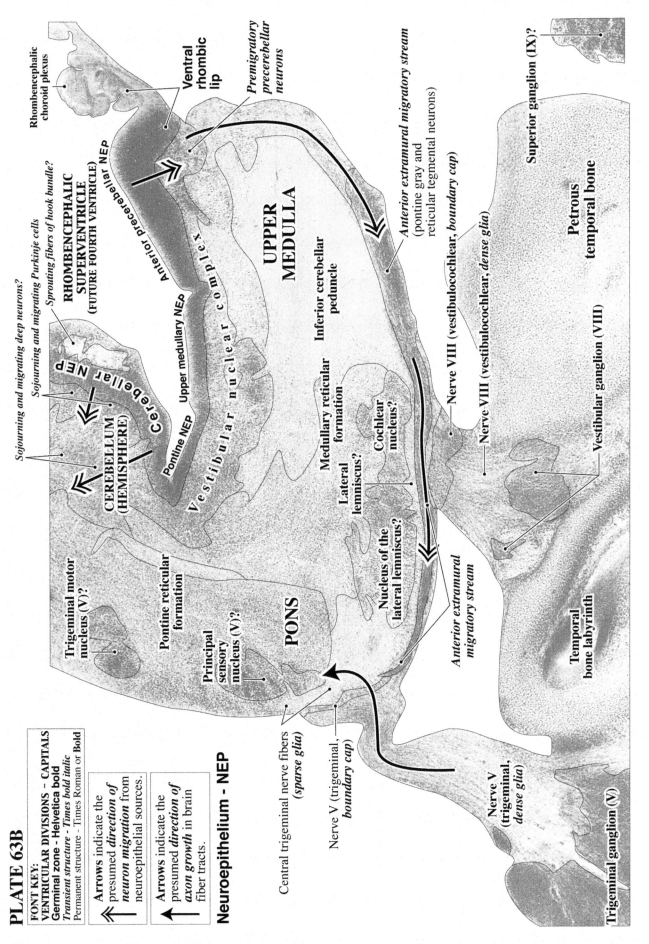

PLATE 63B

FONT KEY:
VENTRICULAR DIVISIONS – CAPITALS
Germinal zone – Helvetica bold
Transient structure – Times bold italic
Permanent structure – Times Roman or Bold

⇐ Arrows indicate the presumed *direction of neuron migration* from neuroepithelial sources.

← Arrows indicate the presumed *direction of axon growth* in brain fiber tracts.

Neuroepithelium – NEP

Rhombencephalic choroid plexus

Sojourning and migrating deep neurons?

Sojourning and migrating Purkinje cells

Sprouting fibers of hook bundle?

Ventral rhombic lip

Premigratory precerebellar neurons

RHOMBENCEPHALIC SUPERVENTRICLE
(FUTURE FOURTH VENTRICLE)

Anterior precerebellar NEP

Cerebellar NEP

CEREBELLUM (HEMISPHERE)

Pontine NEP   Upper medullary NEP

Vestibular nuclear complex

UPPER MEDULLA

Inferior cerebellar peduncle

Anterior extramural migratory stream
(pontine gray and reticular tegmental neurons)

Nerve VIII (vestibulocochlear, *boundary cap*)

Nerve VIII (vestibulocochlear, *dense glia*)

Superior ganglion (IX)?

Petrous temporal bone

Medullary reticular formation

Cochlear nucleus?

Lateral lemniscus?

Nucleus of the lateral lemniscus?

Trigeminal motor nucleus (V)?

Pontine reticular formation

Principal sensory nucleus (V)?

PONS

Anterior extramural migratory stream

Nerve V (trigeminal, *boundary cap*)

Central trigeminal nerve fibers (*sparse glia*)

Nerve V (trigeminal, *dense glia*)

Vestibular ganglion (VIII)

Temporal bone labyrinth

Trigeminal ganglion (V)

170

PLATE 64A

CR 33 mm, GW 9.6, C145,
Sagittal, Slide 27, Section 4

0.5 mm

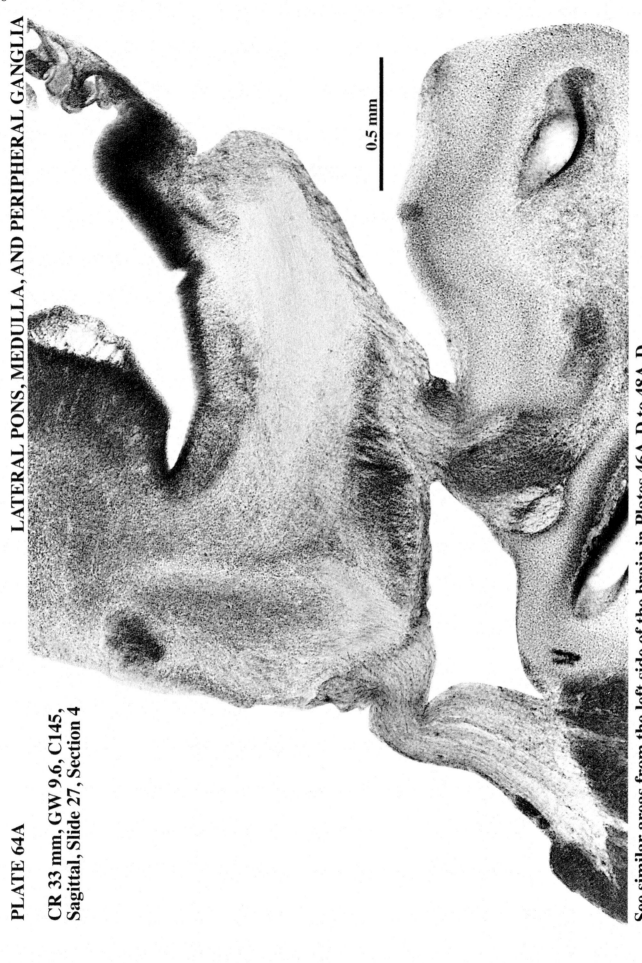

See similar areas from the left side of the brain in Plates 46A-D to 48A-D.

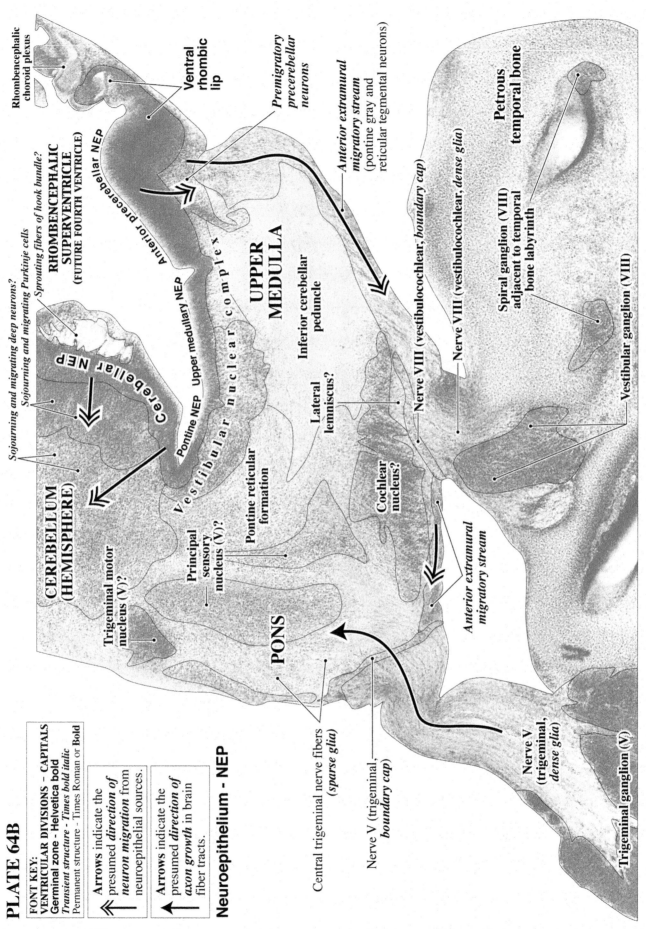

PLATE 64B

FONT KEY:
VENTRICULAR DIVISIONS – CAPITALS
Germinal zone - Helvetica bold
Transient structure - Times bold italic
Permanent structure - Times Roman or Bold

⇐ Arrows indicate the presumed *direction of neuron migration* from neuroepithelial sources.

← Arrows indicate the presumed *direction of axon growth* in brain fiber tracts.

Neuroepithelium - NEP

Rhombencephalic choroid plexus

Ventral rhombic lip

Premigratory precerebellar neurons

Anterior extramural migratory stream (pontine gray and reticular tegmental neurons)

Sojourning and migrating deep neurons?

Sojourning and migrating Purkinje cells

Sprouting fibers of hook bundle?

RHOMBENCEPHALIC SUPERVENTRICLE (FUTURE FOURTH VENTRICLE)

Anterior precerebellar NEP

Cerebellar NEP

CEREBELLUM (HEMISPHERE)

Pontine NEP   Upper medullary NEP

V e s t i b u l a r   n u c l e a r   c o m p l e x

UPPER MEDULLA

Inferior cerebellar peduncle

Petrous temporal bone

Nerve VIII (vestibulocochlear, *dense glia*)

Spiral ganglion (VIII) adjacent to temporal bone labyrinth

Lateral lemniscus?

Nerve VIII (vestibulocochlear, *boundary cap*)

Vestibular ganglion (VIII)

Trigeminal motor nucleus (V)?

Principal sensory nucleus (V)?

Pontine reticular formation

Cochlear nucleus?

PONS

Anterior extramural migratory stream

Central trigeminal nerve fibers (*sparse glia*)

Nerve V (trigeminal, *boundary cap*)

Nerve V (trigeminal, *dense glia*)

Trigeminal ganglion (V)